George Farrar Rodwell

Etna

A History of the Mountain and of its Eruptions

George Farrar Rodwell

Etna

A History of the Mountain and of its Eruptions

ISBN/EAN: 9783743332225

Manufactured in Europe, USA, Canada, Australia, Japa

Cover: Foto ©berggeist007 / pixelio.de

Manufactured and distributed by brebook publishing software
(www.brebook.com)

George Farrar Rodwell

Etna

ETNA

A HISTORY OF THE MOUNTAIN AND OF ITS ERUPTIONS.

BY

G. F. RODWELL,

SCIENCE MASTER IN MARLBOROUGH COLLEGE.

WITH MAPS AND ILLUSTRATIONS.

LONDON

C. KEGAN PAUL & Co., 1, PATERNOSTER SQUARE

1878

I DEDICATE THIS BOOK

TO

MY MOTHER.

PREFACE.

WHILE preparing an account of MOUNT ETNA for the Encyclopædia Britannica, I was surprised to find that there exists no single work in the English language devoted to the history of the most famous volcano in the world. I was consequently induced to considerably enlarge the Encyclopædia article, and the following pages are the result. The facts recorded have been collected from various sources—German, French, Italian, and English, and from my own observations made during the summer of 1877. I desire to express my indebtedness to Mr. Frank Rutley, of H.M. Geological Survey, for his careful examination of the lavas which were collected during my ascent of the mountain, and for the account which he has written of them; also to

Mr. John Murray for permission to copy figures from Lyell's " Principles of Geology."   My thanks are also due to Mr. George Dennis, H.M. Consul-General in Sicily ; Mr. Robert O. Franck, Vice-Consul in Catania ; and to Prof. Orazio Silvestri, for information with which they have severally supplied me.

G. F. RODWELL.

Marlborough,

September 6th, 1878.

CONTENTS.

CHAPTER I.

HISTORY OF THE MOUNTAIN.

CHAPTER II.

PHYSICAL FEATURES OF THE MOUNTAIN.

CHAPTER VI.

GEOLOGY AND MINERALOGY OF THE MOUNTAIN.

LIST OF ILLUSTRATIONS.

ETNA.

A HISTORY OF THE MOUNTAIN AND OF ITS ERUPTIONS.

CHAPTER I.

HISTORY OF THE MOUNTAIN.

THE principal mountain chain of Sicily skirts the
North and a portion of the North-eastern coast, and
would appear to be a prolongation of the Apennines.
An inferior group passes through the centre of the
island, diverging towards the South, as it approaches
the East coast. Between the two ranges, and completely

B

separated from them by the valleys of the Alcantara
and the Simeto, stands the mighty mass of Mount Etna,
which rises in solitary grandeur from the eastern sea-
board of the island. Volcanoes, by the very mode of
their formation, are frequently completely isolated ; and,
if they are of any magnitude, they thus acquire an
imposing contour and a majesty, which larger mountains,
forming parts of a chain, do not possess. This specially
applies to Etna. "Cœlebs degit," says Cardinal Bembo,
"et nullius montis dignata conjugium, caste intra suos
terminos continetur." It is not alone the conspicuous
appearance of the mountain which has made it the
most famous volcano either of ancient or modern times :
—the number and violence of its eruptions, the extent
of its lava streams, its association with antiquity, and
its history prolonged over more than 2400 years, have
all tended to make it celebrated.

The geographical position of Etna was first accurately
determined by Captain Smyth in 1814. He estimated
the latitude of the highest point of the bifid peak of
the great crater at 37° 43′ 31″ N. ; and the longitude
at 15° East of Greenwich. Elie de Beaumont repeated
the observations in 1834 with nearly the same result ;
and these determinations have been very generally
adopted. In the new Italian map recently constructed
by the Stato Maggiore, the latitude of the centre of the
crater is stated to be 37° 44′ 55″ N., and the longitude

44' 55" E. of the meridian of Naples, which passes through the Observatory of Capo di Monte.

According to Bochart the name of Etna is derived from the Phœnician *athana*—a furnace ; others derive it from αἴθω—to burn. Professor Benfey of Gottingen, a great authority on the subject, considers that the word was created by one of the early Indo-Germanic races. He identifies the root *ait* with the Greek αἴθ and the Latin *aed*—to burn, as in *aes*-tu. The Greek name Αἴτνα was known to Hesiod. The more modern name, *Mongibello*, by which the mountain is still commonly known to the Sicilians, is a combination of the Arabic *Gibel*, and the Italian *Monte*. During the Saracenic occupation of Sicily, Etna was called *Gibel Uttamat*—the mountain of fire ; and the last syllables of Mongibello are a relic of the Saracenic name. A mountain near Palermo is still called Gibel Rosso—the red mountain ; and names may not unfrequently be found in the immediate neighbourhood of Etna which are partly, or sometimes even entirely, composed of Arabic words ; such, for example, as *Alcantara*—the river of *the bridge*. Etna is also often spoken of distinctively as *Il Monte*—the mountain *par excellence ;* a name which, in its capacity of the largest mountain in the kingdom of Italy, and the loftiest volcano in Europe, it fully justifies.

Etna is frequently alluded to by classical writers. By the poets it was sometimes feigned to be the prison of the giant Enceladus or Typhon, sometimes the forge of Hephaistos, and the abode of the Cyclops.

It is strange that Homer, who has so minutely described certain portions of the contiguous Sicilian coast, does not allude to Etna. This has been thought by some to be a proof that the mountain was in a quiescent state during the period which preceded and coincided with the time of Homer.

Pindar (B.C. 522–442) is the first writer of antiquity who has described Etna. In the first of the Pythian Odes for Hieron, of the town of Aitna, winner in the chariot race in B.C. 474, he exclaims :

. . . . " He (Typhon) is fast bound by a pillar of the sky, even by snowy Etna, nursing the whole year's ength her dazzling snow. Whereout pure springs of unapproachable fire are vomited from the inmost depths : in the daytime the lava-streams pour forth a lurid rush of smoke ; but in the darkness a red rolling flame sweepeth rocks with uproar to the wide deep sea . . . That dragon-thing (Typhon) it is that maketh issue from beneath the terrible fiery flood." [1]

Æschylus (B.C. 525–456) speaks also of the " mighty Typhon," (*Prometheus* V.) :

> . . . . . " He lies
> A helpless, powerless carcase, near the strait
> Of the great sea, fast pressed beneath the roots

---

[1] Translated by Ernest Myers, M.A., 1874.

Of ancient Etna, where on highest peak
Hephæstos sits and smites his iron red hot,
From whence hereafter streams of fire shall burst,
Devouring with fierce jaws the golden plains
Of fruitful, fair Sikelia." [1]

Herein he probably refers to the eruption which had occurred a few years previously (B.C. 476).

Thucydides (B.C. 471–402) alludes in the last lines of the Third Book to several early eruptions of the mountain in the following terms : " In the first days of this spring, the stream of fire issued from Etna, as on former occasions, and destroyed some land of the Catanians, who live upon Mount Etna, which is the largest mountain in Sicily. Fifty years, it is said, had elapsed since the last eruption, there having been three in all since the Hellenes have inhabited Sicily." [2]

Virgil's oft-quoted description of the mountain (*Eneid*, Bk. 3) we give in the spirited translation of Conington :

" But Etna with her voice of fear
In weltering chaos thunders near.
Now pitchy clouds she belches forth
Of cinders red, and vapour swarth ;
And from her caverns lifts on high
Live balls of flame that lick the sky :
Now with more dire convulsion flings
Disploded rocks, her heart's rent strings,
And lava torrents hurls to-day
A burning gulf of fiery spray."

[1] Translated by E. Myers.
[2] Translated by E. Crawley.

Many other early writers speak of the mountain, among them Theokritos, Aristotle, Ovid, Livy, Seneca, Lucretius, Pliny, Lucan, Petronius, Cornelius Severus, Dion Cassius, Strabo, Diodorus Siculus, and Lucilius Junior. Seneca makes various allusions to Etna, and mentions the fact that lightning sometimes proceeded from its smoke.

Strabo has given a very fair description of the mountain. He asserts that in his time the upper part of it was bare, and covered with ashes, and in winter with snow, while the lower slopes were clothed with forests. The summit was a plain about twenty stadia in circumference, surrounded by a ridge, within which there was a small hillock, the smoke from which ascended to a considerable height. He further mentions a second crater. Etna was commonly ascended in Strabo's time from the south-west.

While the poets on the one hand had invested the mountain with various supernatural attributes, and had made it the prison-house of a chained giant, and the workshop of a swart god, Lucretius endeavoured to show that the eruptions and other phenomena could be easily explained by the ordinary operations of nature. "And now at last," he writes, "I will explain in what ways yon flame, roused to fury in a moment, blazes forth from the huge furnaces of Aetna. And, first, the nature of the whole mountain is hollow underneath, underpropped throughout with caverns of basalt rocks. Furthermore, in all caves are wind and air, for wind is produced when the air has been stirred and put in motion. When this air has been thoroughly heated, and, raging about, has imparted its heat to all the rocks around, wherever it comes in contact with them, and to the earth, and has struck out from them fire burning with swift flames, it rises up and then forces itself out on high, straight through the gorges ; and so carries its heat far, and scatters far its ashes, and rolls on smoke of a thick pitchy blackness, and flings out at the same time stones of prodigious weight—leaving no doubt that this is the stormy force of air. Again, the sea, to a great extent, breaks its waves and sucks back its surf at the roots of that mountain. Caverns reach from this sea as far as the deep gorges of the mountain below. Through these you must admit [that air mixed up with water passes ; and] the nature of the case compels [this air to enter in from that] open sea, and pass right within, and then go out in blasts, and so lift up flame, and throw out stones, and raise clouds of sand ; for on the summit are craters, as they name them in their own language, what we call gorges and mouths." [1]

These ideas were developed by Lucilius Junior in a poem consisting of 644 hexameters entitled *Aetna*. The

---

[1] *De Naturâ Rerum*, Book 6, p. 580. Translated by E. Munro 1864.

authorship of this poem has long been a disputed point ;
it has been attributed to Virgil, Claudian, Quintilius
Varus, Manilius, and, by Joseph Scaliger [1] and others,
to Cornelius Severus. Wensdorff was the first to adduce
reasons for attributing the poem to Lucilius Junior, and
his views are generally adopted. Lucilius Junior was
Procurator of Sicily under Nero, and, while resident in
the Island, he ascended Etna ; and it is said that he
proposed writing a detailed history of the mountain.
He adopted the scientific opinions of Epicurus, as established in Rome by Lucretius, and was more immediately a disciple of Seneca. The latter dedicated to
him his *Quæstiones Naturales*, in which he alludes more
than once to Etna. M. Chenu speaks of the poem of
Lucilius Junior as "sans doute très-poétique, mais
assez souvent dur, heurté, concis, et parcela même, d'une
obscurité parfois désespérante."[2] At the commencement
of the poem, Lucilius ridicules the ideas of the poets as
regards the connection of Etna with Vulcan and the
Cyclops. He has no belief in the practice, which
apparently prevailed in his time, of ascending to the
edge of the crater and there offering incense to the

---

[1] See Lucilius Junioris AETNA. Recensuit notasque Jos.
Scaligeri, Frid. Lindenbruchii et suas addidit Fridericus Jacob.
Lipsiæ, 1826.

[2] L'Etna de Lucilius Junior. Traduction nouvelle par Jules
Chenu. Paris, 1843.

tutelary gods of the mountain. He adopts to a great extent the tone and style of Lucretius, in his explanation of the phenomena of the mountain. Water filters through the crevices and cracks in the rocks, until it comes into contact with the internal fires, when it is converted into vapour and expelled with violence. The internal fires are nourished by the winds which penetrate into the mountain. He traces some curious connection between the plants which grow upon the mountain, and the supply of sulphur and bitumen to the interior, which is, at best, but partly intelligible.

> " Nunc superant, quacunque regant incendia silvæ
> Quæ flammis alimenta vacent, quid nutriat Aetnam.
> Incendi patiens illis vernacula caulis
> Materia, appositumque igni genus utile terræ est,
> Uritur assidue calidus nunc sulfuris humor,
> Nunc spissus crebro præbetur flumine succus,
> Pingue bitumen adest, et quidquid cominus acres
> Irritat flammas; illius corporis Ætna est.
> Atque hanc materiam penitus discurrere fontes
> Infectæ erumpunt et aquæ radice sub ipsa."

Many of the myths developed by the earlier poets had their home in the immediate neighbourhood, sometimes upon the very sides, of Etna—Demeter seeking Persephone; Acis and Galatea; Polyphemus and the Cyclops. Mr. Symonds tells us that the one-eyed giant Polyphemus was Etna itself, with its one great crater, while the Cyclops were the many minor cones. "Per-

sephone was the patroness of Sicily, because amid the
billowy corn-fields of her mother Demeter, and the
meadow-flowers she loved in girlhood, are ever found
sulphurous ravines, and chasms breathing vapour from
the pit of Hades." [1]

It is said that both Plato and the Emperor Hadrian
ascended Etna in order to witness the sunrise from its
summit. The story of

> " He who to be deemed
> A god, leaped fondly into Etna flames,
> Empedokles "

is too trite to need repetition. A ruined tower near the
head of the Val del Bove, 9,570 feet above the sea, has
from time immemorial been called the *Torre del Filosofo*,
and is asserted to have been the observatory of Em-
pedokles. Others regard it as the remains of a Roman
tower, which was possibly erected on the occasion of
Hadrian's ascent of the mountain.

During the Middle Ages Etna is frequently alluded
to, among others by Dante, Petrarch, Boccaccio, and
Cardinal Bembo. The latter gives a description of the
mountain in the form of a dialogue, which Ferrara
characterises as *"erudito, e grecizzante, ma sensa nervi."*
He describes its general appearance, its well-wooded
sides, and sterile summit. When he visited the
mountain it had two craters about a stone's throw

---

[1] Sketches in Italy and Greece, p. 201.

apart; the larger of the two was said to be about three miles in circumference, and it stood somewhat above the other.[1]

In 1541 Fazzello made an ascent of the mountain, which he briefly describes in the fourth chapter of his bulky volume *De Rebus Siculis*.[2] This chapter is entitled *"De Aetna monte et ejus ignibus;"* it contains a short history of the mountain, and some mention of the principal towns which he enumerates in the following order: Catana, Tauromenium, Caltabianco, Linguagrossa, Castroleone, Francavilla, Rocella, Randatio (Randazzo), Bronte, Adrano, Paterno, and Motta. Fazzello speaks of only one crater.

In 1591 Antonio Filoteo, who was born on Etna, published a work in Venice in which he describes the general features of the mountain, and gives a special account of an eruption which he witnessed in 1536.[3] The mountain was then, as now, divided into three *Regions*. The first and uppermost of these, he asserts, is very arid, rugged, and uneven, and full of broken

---

[1] Petri Bembi DE AETNA. Ad Angelum Chabrielem Liber Impressum Venetiis Aedibus Aldi Romani. Mense Februario anno M.V.D. (1495).

[2] Fazzellus T. DE REBUS SICULIS. Panormi, 1558; folio.

[3] Antonii Philothei de Homodeis Siculi, AETNÆ TOPOGRAPHIA, incendiorum Aetnæorum Historia. Venetiis. 1591. Preface dated September, 1590.

rocks; the second is covered with forests; and the third is cultivated in the ordinary manner. Of the height he remarks, "Ascensum triginta circiter millia passuum ad plus habet." In regard to the name, *Mongibello*, he makes a curious error, deriving it from *Mulciber*, one of the names of Vulcan, who, as we have seen, was feigned by the earlier poets to have had his forge within the mountain.

In 1636 Carrera gave an account of Etna, followed by that of the Jesuit Kircher, in 1638. The great eruption of 1669 was described at length by various eye-witnesses, and furnished the subject of the first detailed description of the eruptive phenomena of the mountain. Public attention was now very generally drawn to the subject in all civilised countries. It was described by the naturalist, Borelli, and in our own *Philosophical Transactions*. Lord Winchelsea, our ambassador at Constantinople, was returning to England by way of the Straits of Messina at the time of the eruption, and he forwarded to Charles II "A true and exact relation of the late prodigious earthquake and eruption of Mount Ætna, or Monte Gibello."

The first map of the mountain which we have been able to meet with, was published in reference to the eruption of 1669; it is entitled, "Plan du Mont Etna communcnent dit Mount Gibel en l'Isle de Scicille et de t'jncëdie arrive par un trëblement de terre le 8me

Mars dernier 1669." This plan is in the Bibliothèque Nationale, in Paris; it was probably drawn from a simple description, or perhaps altogether from the imagination, as it is utterly unlike the mountain, the sides of which possess an impossible steepness. Another very inaccurate map was published in Nuremberg about 1680, annexed to a map of Sicily, which is entitled, "*Regnorum Siciliæ et Sardiniæ, Nova Tabula.*" Again, in 1714 H. Moll, "geographer in Devereux Street, Strand," published a new map of Italy, in which there is a representation of Etna during the eruption of 1669. This also was probably drawn from the imagination; no one who has ever seen the mountain would recognise it, for it has a small base, and sides which rival the Matterhorn in abruptness. Over against the coast of Sicily, and near the mountain, is written:—
"Mount Etna, or Mount Gibello. This mountain sometimes issues out pure flame, and at other times a thick smoak with ashes; streams of fire run down with great quantities of burning stones, and has made many eruptions."

During the eighteenth century Etna was frequently ascended, and as frequently described. We have the accounts of Massa (1703), Count D'Orville (1727), Riedesel (1767), Sir William Hamilton (1769), Brydone (1776), Houel (1786), Dolomieu (1788), Spallanzani (1790), and many minor writers, such as Borch, Brocchi,

Swinburne, Denon, and Faujas de Saint Fond. There is great sameness in all of these narratives, and much repetition of the same facts; some of them, however, merit a passing notice.

Sir William Hamilton's *Campi Phlegræi* relates mainly to Vesuvius and the surrounding neighbourhood; but one of the letters "addressed to the Secretary of the Royal Society on October 17th, 1769," describes an ascent of Etna. Hamilton ascended on June 24th with the Canon Recupero and other companions; the few observations of any value which he made have been alluded to elsewhere under the head of the special subjects to which they refer. The illustrations of the *Campi Phlegræi*, specially the original water-colours which are contained in one of the British Museum copies, are magnificent, and convey a better idea of volcanic phenomena than any amount of simple reading. From them we can well realise the opening of a long rift extending down the sides of a mountain during its eruption, and the formation of subsidiary craters along the line of fire thus opened. Various volcanic products are also admirably painted. In the picture of Etna, however, which was drawn by Antonio Fabris, the artist has scarcely been more successful than his predecessors, and the slope of the sides of the mountain has been greatly exaggerated.

M. Houel, in his *Voyage pittoresque dans les Deux*

Siciles, 1781–1786, has given a fairly good account of Etna, accompanied by some really excellent engravings.

In 1776 Patrick Brydone, a clever Irishman with a good deal of native shrewdness and humour, published two volumes of a *Tour in Sicily and Malta*, in which he devoted several chapters to Mount Etna. He made the ascent of the mountain, and collected from the Canon Recupero, and from others, many facts concerning its then present, and its past history. He also made observations as to the height, temperature of the air at various elevations, brightness of the stars, and so on. Sir William Hamilton calls Brydone "a very ingenious and accurate observer," and adds that he was well acquainted with Alpine measurements. M. Elie de Beaumont, writing in 1836, speaks of him as *le celebre Brydone;* while, on the other hand, the Abbé Spallanzani, displeased at certain remarks which he made concerning Roman Catholicism in Sicily, never fails to deprecate his work, and deplores "his trivial and insipid pleasantries." Albeit Brydone's chapters on Etna furnished a more complete account of the volcano than any which had appeared in English up to that time; his remarks are frequently very sound and just, and we shall have occasion more than once to quote him.

It was reserved, however, for the Abate Francesco Ferrara, Professor of Physical Science in the University

of Catania, to furnish the first history of Etna and of
its eruptions, which had any just claim to completeness.
It is entitled, *Descrizione dell' Etna, con la Storia delle
Eruzioni e il Catalogo dei prodotti.* The first edition appeared in 1793, and a second was printed in Palermo in
1818. The author had an enthusiastic love for his subject :—" Nato sopra l'Etna," he writes, " che io conobbi
ben presto palmo a palmo la mia passione per lo studio
fissò la mia attenzione sul bello, e terribile fenomeno che
avea avanti agli occhi." The work commences with a
general description of the mountain—its height, the
temperature of the different regions, the view from the
summit, the mass, the water-springs, the vegetable and
animal life, and the internal fires. This extends over
sixty-nine octavo pages. The second part of the book—
eighty pages—gives a history of the eruptions from the
earliest times to the year 1811 ; the third part—sixty-
seven pages—treats of the nature of the volcanic
products ; and the fourth part—thirty-four pages—dis-
cusses certain geological and physical considerations
concerning the mountain. At the end there are a few
badly drawn and engraved woodcuts, and a map which,
although the trend of the coast-line is quite wrong, is
otherwise fairly good. The engravings represent the
mountain as seen from Catania ; the Isole dei Ciclopi,
and the neighbouring coast ; the Montagna della Motta ;
and a view from Catania of the eruption of 1787. This

work has evidently to a great extent been a labour of love; it is full of personal observations, and also embodies the results of many other observers. It has furnished the foundation of much that has since been written concerning Etna.

The Canon Recupero has been alluded to above; he accompanied Hamilton, Brydone, and others to the summit of the mountain, and he was employed by the Government to report on the flood which, in 1755, descended with extraordinary violence through the Val del Bove. Beyond this, Recupero does not appear to have published anything concerning Etna, although it was well known that he had plenty of materials. He died in 1778, and it was not till the year 1815 that his results were published under the title of *Storia Naturale et Generale dell' Etna, del Canonico Guiseppe Recupero.— Opera Postuma.* This work consists of two bulky quarto volumes, the first of which is devoted to a general description of the mountain, the second to a history of the eruptions, and an account of the products of eruptions. Some idea may be formed of the extreme prolixity of the author if we mention that two chapters, together containing twelve quarto pages, are devoted to the discussion of the height of Etna, while the first volume is terminated by sixty-three closely printed pages of annotations. A few rough woodcuts accompany the volumes; a view of the mountain which, as usual, is out of all reason as regards abruptness of ascent, and a *carta oryctographia di Mongibello* in which the trend of the coast-line between Catania and Taormina is altogether inexact, complete the illustrations of this most detailed of histories.

During the years 1814—1816 Captain Smyth, acting under orders from the Admiralty, made a survey of the coast of Sicily, and of the adjacent islands. At this time the Mediterranean charts were very defective; some places on the coast of Sicily were mapped as much as twenty miles out of their true position, and even the exact positions of the observatories at Naples, Palermo, and Malta were not known. Among other results, Smyth carefully determined the latitude and longitude of Etna, accurately measured its height, and examined the surroundings of the mountain. His results were published in 1824, and are often regarded as the most accurate that we possess.

In 1824 Dr. Joseph Gemellaro, who lived all his life on the mountain, and made it his constant study, published an " Historical and Topographical Map of the Eruptions of Etna from the era of the Sicani to the year 1824." In it he delineates the extent of the three Regions, *Coltivata, Selvosa,* and **Deserta**; he places the minor cones, to the number of seventy-four, in their proper places, and he traces the course of the various lava-streams which have flowed from them and from the great crater. This map is the result of much patient labour and study, and it is a great improvement upon those of Ferrara and Recupero, but of course it is impossible for one man to survey with much accuracy an area of nearly 500 square miles, and to trace the tortuous course of a large number of lava-streams. Hence we must be prepared for inaccuracies, and they are not uncommon—the coast line is altogether wrong as to its bearings, some of the small towns on the sides of the mountain are misplaced, and but little attention has been paid to scale. Still the map is very useful, as it is the only one which shows the course of the lava-streams.

Mario Gemellaro, brother of the preceding, made almost daily observations of the condition of Etna, between the years 1803 and 1832. These results were tabulated, and they are given in the *Vulcanologia dell' Etna* of his brother, Professor Carlo Gemellaro, under the title of *Registro di Osservazioni del Sigr. Mario Gemellaro.*

Carlo Gemellaro contributed many memoirs on subjects connected with the mountain. They are chiefly to be found in the *Atti dell' Accademia Gioenia* of Catania, and they extend over a number of years. Perhaps the most important is the treatise entitled "*La Vulcanologia dell' Etna che comprende la Topographia, la Geologia, la Storia, delle sue Eruzioni.*" It was published in Catania in 1858, and is dedicated to Sir Charles Lyell, who ascended the mountain under the guidance of the author. The latter also published a *Breve Raggualio della Eruzione dell' Etna, del* 21 *Agosto* 1852, which contains the most authentic account of this important eruption, accompanied by some graphic sketches made on the spot. The last contribution of Carlo Gemellaro to the history of Etna, is fitly entitled *Un Addio al Maggior Vulcano di Europa.* It was published in 1866, and with pardonable vanity the author reviews his work in connection with the mountain, extending over a period of more than forty years. He commences his somewhat florid farewell with the following apostrophe :—" O Etna ! splendida e perenne manifestazione della esistenza dei Fuochi sotteranei massimo fra quanti altri monti, dalle coste meridionali di Europa, dalle orientali dell' Asia e delle settentrionali dell' Africa si specchiano nel Mediterraneo: tremendo pei tuoi incendii : benigno per la fertilità del vulcanico tuo terreno ridotto a prospera coltivazione . . . io, nato appiè del vasto tuo cono, in quella Città che hai minacciato più d'una volta di sepellire sotto le tue infocate correnti : allogato, nella mia prima età, in una stanza della casa paterna, che signoreggiava in allora più basse abizioni vicine, ed intiera godeva la veduta della estesa parte meridionalè della tua mole, io non potera non averti di continuo sotto gli occhi, e non essere spettatore dei tuoi visibili fenomeni ! "

In 1834, M. Elie de Beaumont commenced a minute geological examination of the mountain. His results were published in 1838, under the title of *Recherches sur la structure et sur l'origine du Mont Etna*, and they extend over 225 pages.[1] He re-determined the latitude and longitude of the mountain, measured the slope of the cone, and the diameter of the great crater, and minutely examined the structure of the rocks at the base of the mountain. He also gives a good sectional view, elevations taken from each quarter of the compass, and a geological map, which although accurate in its general details, can scarcely be considered very satisfactory. A relief map of Etna, a copy of which is in the Royal School of Mines, was afterwards constructed from the flat map, and this was, we believe, at the same time, the first geological map, and the first map in relief, which had been made of the mountain. Elie de Beaumont considers granite as the basis of the mountain, because it is sometimes ejected from the crater; old basaltic rocks appear in the Isole dei Ciclopi, and near Paterno, Licodia, and Aderno; *cailloux roulés* near Motta; ancient lavas on each side of the Val del Bove; modern lavas in every part of the mountain, and calcareous and arenaceous rocks in the surrounding mountains.

[1] Printed in vol. IV. of *Mémoires pour servir à une description Géologique de la France*. Par M.M. Dufrénoy et Elie de Beaumont.

In 1836, Abich published some excellent sections of Etna, and an accurate view of the interior of the crater, in a work entitled *Vues illustratives de quelques Phénomènes Géologiques. prises sur le Vésure et l'Etna pendant les années* 1833 *and* 1834.

The whole of the thirteenth volume (1839) of the Berlin *Archiv für Mineralogie, Geognosie, Berghau und Hüttenkunde,* is occupied by an elaborate memoir on the geology of Sicily [1] by Friedrich Hoffmann, accompanied by an excellent geological map. A long account of the geology of Etna is given, and an enlarged map of the mountain was afterwards constructed and published in the *Vulkanen Atlas* of Dr. Leonhard in 1850.[2]

In 1836 Baron Sartorius Von Waltershausen commenced a minute survey of Etna, preparatory to a complete description of the mountain, both geological and otherwise. He was assisted by Professor Cavallari of Palermo, Professor Peters of Hensbourg, and Professor Roos of Mayence. The survey occupied six years, (1836–1842), and the results of direct observation in the form of maps and drawings, occupied a hundred sheets 160 millimetres (6¼ inches) long, by 133 m.m.

---

[1] Entitled *Geognostiche Beobachtungen Gesammelt auf einer Reise durch Italian und Sicilien, in den jahren* 1830 *bis* 1832, *von Friedrich Hoffmann.*

[2] *Vulkanen Atlas zur naturgeschichte der Erde von K. G. Von Leonhard.* Stuttgardt. 1850.

(5¼ inches) broad. Twenty-nine separate points were made use of in the triangulation; and the scale chosen was 1 in 50,000. The results were published in a large folio atlas, which appeared in eight parts; the first in 1845 and the last in 1861, when the death of Von Waltershausen put an end to the further publication. There are 26 fine coloured maps, and 31 engravings. The cost of the atlas is £12. The maps are both geological and topographical, and they are accompanied by outline engravings of various details of special interest. The *Atlas des Aetna* furnishes the most exhaustive history of any one mountain on the face of the earth, and Sartorius Von Waltershausen will always be the principal authority on the subject of Etna.

Sir Charles Lyell visited Etna in 1824, 1857, and again in 1858. He embodied his researches in a paper presented to the Royal Society in 1859, and in a lengthy chapter in the *Principles of Geology*. His investigations have added much to our knowledge of the formation and geological characteristics of the mountain, especially of that part of it called the Val del Bove.

Later writers usually quote Von Waltershausen and Lyell, and do not add much original matter. The facts of all subsequent writers are taken more or less directly from these authors. The latest addition to the literature of the mountain, is the *Wanderungen am Aetna* of

Dr. Baltzer, in the journal of the Swiss Alpine Club for 1874. [1]

A fine map of Sicily, on the unusually large scale of 1 in 50,000, or 1·266 inch to a mile, was constructed by the Stato Maggiore of the Italian government, between 1864 and 1868. The portion relating to Etna, and its immediate surroundings occupies four sheets. All the small roads and rivulets are introduced; the minor cones and monticules are placed in their proper positions, and the elevation of the ground is given at short intervals of space over the entire map. An examination of this map shows us that distances, areas, and heights, have been repeatedly misstated, the minor cones misplaced, and the trend of the coast line misrepresented. For example, if we draw a line due north and south through Catania, and a second line from the Capo di Taormina, (the north-eastern limit of the base of Etna), until it meets the first line at Catania, the lines will be found to enclose an angle of 26°. If we adopt the same plan with Gemellaro's map, the included angle is found to be 53°, and in the case of the maps of Ferrara and Recupero more than 60°. Again, it has been stated on good authority, that the lava of 396 B.C. which enters the sea at Capo di Schiso flowed for a distance of nearly 30 miles; the map shows us that its true course was less than 16 miles.

---

[1] Jahrbuch des Schweizer Alpen Club. Neunter Jahrgang, 1873—1874. Bern 1874.

Lyell in 1858 gives a section of the mountain from West 20° N., to East 20° S., but a comparison with the new map proves that the section is really taken from West 35° N. to East 35° S., an error which at a radius of ten miles from the crater would amount to a difference of nearly three miles.

The mantle of Carlo Gemellaro appears to have fallen upon Cav. Orazio Silvestri, Professor of Chemistry in the University of Catania. He has devoted himself with unwearying vigour to the study of the mountain, and his memoirs have done much to elucidate its past and present history. His most recent work of importance on the subject is entitled *I Fenomeni Vulcanici presentati dall' Etna nel 1863-64-65-66.* It was published in Catania in 1867, and contains an account of some very elaborate chemico-geological researches.

CHAPTER II.

PHYSICAL FEATURES OF THE MOUNTAIN.

Height.—Radius of Vision from the summit.—Boundaries.—
Area.—Population.—General aspect of Etna.—The Val del Bove.
—Minor Cones.—Caverns.—Position and extent of the three
Regions. — Regione Coltivata. — Regione Selvosa. — Regione
Deserta. — Botanical Regions. — Divisions of Rafinesque-
Schmaltz, and of Presl.—Animal life in the upper Regions.

In the preceding chapter we have discussed the
history of Mount Etna; the references to its phenomena
afforded by writers of various periods; and the present
state of the literature of the subject. We have now to
consider the general aspect and physical features of the
mountain, together with the divisions of its surface into
distinct regions.

The height of Etna has been often determined. The
earlier writers had very extravagant notions on the
subject, and three miles has sometimes been assigned
to it. Brydone, Saussure, Shuckburgh, Irvine, and
others, obtained approximations to the real height;
it must be borne in mind, however, that the cone of a
volcano is liable to variations in height at different

periods, and a diminution of as much as 300 feet occurred during one of the eruptions of Etna, owing to the falling in of the upper portion of the crater. During the last sixty years, however, the height of the mountain has been practically constant. In 1815 Captain Smyth determined it to be 10,874 feet. In 1824 Sir John Herschel, who was unacquainted with Smyth's results, estimated it at 10,872½ feet. The new map of the Stato Maggiore gives 3312·61 metres=10867·94 feet.

When the Canon Recupero devoted two chapters of his quarto volume to a discussion of the height of Etna, no such exact observations had been made, consequently he compared, and critically examined, the various determinations which then existed. The almost perfect concordance of the results given by Smyth, Herschel, and the Stato Maggiore, render it unnecessary for us to further discuss a subject about which there can now be no difference of opinion.

Professor Jukes says, " If we were to put Snowdon, the highest mountain in Wales, on the top of Ben Nevis, the highest in Scotland, and Carrantuohill, the highest in Ireland, on the summit of both, we should make a mountain but a very little higher than Etna, and we should require to heap up a great number of other mountains round the flanks of our new one in order to build a gentle sloping pile which should equal Etna in bulk.

The extent of radius of vision from the summit of Etna
is very variously stated. The exaggerated notions of the
earlier writers, that the coast of Africa and of Greece are
sometimes visible, may be at once set aside. Lord
Ormonde's statement that he saw the Gulf of Taranta,
and the mountains of Terra di Lecce beyond it—a distance
of 245 miles—must be received with caution. It is,
however, a fact that Malta, 130 miles distant, is often
visible; and Captain Smyth asserts that a considerable
portion of the upper part of the mountain may sometimes
be seen, and that he once saw more than half of it, from
Malta, although that island is usually surrounded by a
sea-horizon. It is stated on good authority that Monte
S. Giuliano above Trapani, and the Œgadean Isles, 160
miles distant, are sometimes seen. Other writers give
128 miles as the limit. The fact is, that atmospheric
refraction varies so much with different conditions of the
atmosphere that it is almost impossible to give any exact
statement. The more so when we remember that there
may be many layers of atmosphere of different density
between the observer and the horizon. Distant objects
seem to be just under one's feet when seen from the summit of the mountain. Smyth gives the radius of vision
as 150·7 miles: and this we are inclined to adopt as the
nearest approach to the truth, because Smyth was an
accurate observer, and he made careful corrections both
for error of instruments and for refraction. This radius

gives an horizon of 946·4 miles of circumference, and an included area of 39,900 square miles—larger than the area of Ireland. If a circle be traced with the crater of Etna as a centre, and a radius of 150·7 miles, it will be found to take in the whole of Sicily and Malta, to cut the western coast of Italy at Scalea in Calabria, leaving the south-east coast near Cape Rizzuto. Such a circle will include the whole of Ireland, or if we take Derby as the centre, its circumference will touch the sea beyond Yarmouth on the East, the Isle of Wight on the South, the Irish Channel on the West, and it will pass beyond Carlisle and Newcastle-on-Tyne on the North.

The road which surrounds the mountain is carried along its lower slopes, and is 87 miles in length. It passes through the towns of Paterno, Aderno, Bronte, Randazzo, Linguaglossa, Giarre, and Aci Reale. It is considered by some writers to define the base of the mountain, which is hence most erroneously said to have a circumference of 87 miles; but the road frequently passes over high beds of lava, and winds considerably. It is about 10 miles from the crater on the North, East, and West sides, increasing to 15½ miles at Paterno, (S.W.) The elevation on the North and West flanks of the mountain is nearly 2,500 feet, while on the South it falls to 1,500 feet, and on the East to within 50 feet of the level of the sea. It is quite clear that it cannot be asserted with any degree of accuracy to define the base of the mountain.

The "natural boundaries" of Etna are the rivers Alcantara and Simeto on the North, West, and South, and the sea on the East to the extent of 23 miles of coast, along which lava streams have been traced, sometimes forming headlands several hundred feet in height. The base of the mountain, as defined by these natural boundaries, is said to have a circumference of " at least 120 miles," an examination of the new map, however, proves that this is over-estimated.

If we take the sea as the eastern boundary, the river Alcantara, (immediately beyond which Monte di Mojo, the most northerly minor cone of Etna is situated), as the northern boundary, and the river Simeto as the boundary on the west and south, we obtain a circumference of 91 miles for the base of Etna. In this estimate the small sinuosities of the river have been neglected, and the southern circuit has been completed by drawing a line from near Paterno to Catania, because the Simeto runs for the last few miles of its course through the plain of Catania, quite beyond the most southerly stream of lava. The Simeto (anciently *Simæthus*) is called the Giaretta along the last three miles of its course, after its junction with the Gurna Longa.

The area of the region enclosed by these boundaries is approximately 480 square miles. Reclus gives the area

Fig.1

Section of Mount Etna in a direction nearly East and West. (Von Abich)

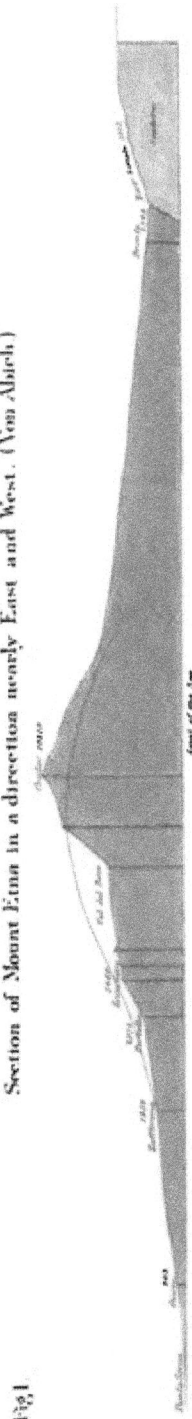

Fig.II.

Section in a direction N.N.W and S.S.E.

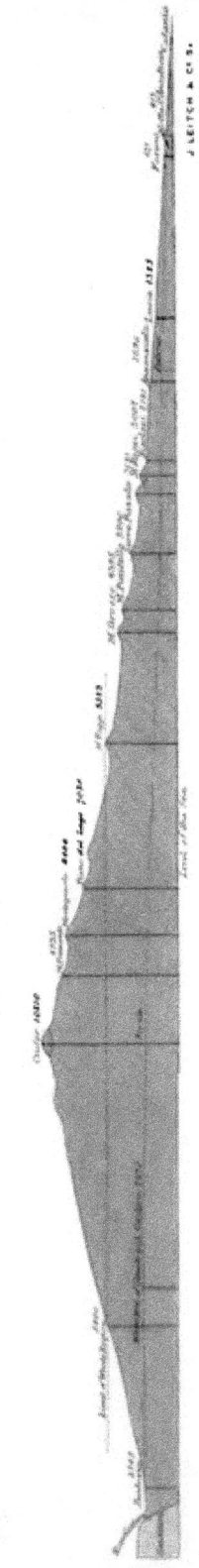

SECTIONS OF ETNA

The Scale is in Paris Feet.

of the mountain as 1,200 square kilometres—461 square miles. (*Nouvelle Geographie Universelle*, 1875.) The last edition of a standard Gazetteer states it as "849 square miles;" but this estimate is altogether absurd. This would require a circle having a radius of between sixteen and seventeen miles. If a circle be drawn with a radius of sixteen miles from the crater, it will pass out to sea to a distance of 4½ miles on the East, while on the West and North it will pass through limestone and sandstone formations far beyond the Alcantara and the Simeto, and beyond the limit of the lava streams.

There are two cities, Catania and Aci Reale, and sixty-two towns or villages on Mount Etna. It is far more thickly populated than any other part of Sicily or Italy, for while the population of the former is 228 per square mile, and of the latter 233, the population of the habitable zone of Etna amounts to 1,424 per square mile. More than 300,000 persons live on the slopes of the mountain. Thus with an area rather larger than that of Bedfordshire (462 square miles) the mountain has more than double the population; and with an area equal to about one-third that of Wiltshire, the population of the mountain is greater by nearly 50,000 inhabitants. We have stated above that the area of Etna is 480 square miles, but it must be borne in mind that the habitable zone only commences at a distance of about 9½ miles from the crater. A circle, having a radius of $9\frac{1}{4}$ miles, encloses an area of 269 square miles; and 480 minus 269 leaves 211 square miles as the approximate area of the habitable zone. Only a few insignificant villages on the East side are nearer to the crater than $9\frac{1}{4}$ miles. Taking the inhabitants as 300,000, we find, by dividing this number by 211, (the area of the habitable zone), that the population amounts to 1,424 per square mile. Even Lancashire, the most populous county in Great Britain, (of course excepting Middlesex), and the possessor of two cities, which alone furnish more than a million inhabitants, has a population of only 1,479 to the square mile.

Some idea of the closeness of the towns and villages may be found by examining the south-east corner of the map. If we draw a line from Aci Reale to Nicolosi, and from Nicolosi to Catania, we enclose a nearly equilateral triangle, having the coast line between Aci Reale and Catania as its third side.

Starting from Aci Reale with 24,151 inhabitants, and moving westwards to Nicolosi, we come in succession to Aci S. Lucia, Aci Catena, Aci S. Antonio, Via Grande, Tre Castagni, Pedara, Nicolosi, completing the first side of the triangle; then turning to the south-east and following the Catania road, we pass Torre di Grifo, Mascalucia, Gravina, and reach Catania with 85,055 inhabitants; while on the line of coast between Catania and Aci Reale we have Ognina, Aci Castello, and Trezza. Within the triangle we find Aci Patane, Aci S. Filippo, Valverde, Bonacorsi, S. Gregorio, Tremestieri, Piano, S. Agata, Trappeto, and S. Giovanni la Punta : in all twenty-five, two of which are cities, several of the others towns of about 3,000 inhabitants, and the rest villages. These are all included within an area of less than thirty square miles, which constitutes the most populous portion f the habitable zone of Etna.

That the population is rapidly increasing is well shown by a comparison of the number of inhabitants of some of the more important towns in 1824 and in 1876.[1]

	1824	*1876*
Catania	45,081	85,055
Aci Reale	14,094	24,151
Giarre	13,705	17,965
Paternò	9,808	16,512
Aderno	6,623	15,657
Bronte	9,153	15,081
Biancavilla	5,870	13,261
Linguaglossa	2,415	9,120
Randazzo	4,700	8,378
Piedimonte Etnea	1,404	4,924
Zaffarana Etnea	700	3,884
Pedara	2,068	3,181
Trecastagni	2,406	3,061

[1] I am indebted for these figures to Mr. George Dennis, H.M. Consul General for Sicily.

The general aspect of Etna is that of a pretty regular cone, covered with vegetation, **except** near the summit. The regularity is broken on the East side by a slightly oval valley, four or five miles in diameter, called the *Val del Bove*, or in the language of the district Val del *Bue*. This commences about two miles from the summit, and is bounded on three sides by nearly vertical precipices from 3,000 to 4,000 feet in height. The bottom of the valley is covered with lavas of various date, and several minor craters have from time to time been upraised from it. Many eruptions have commenced in the immediate neighbourhood of the Val del Bove, and Lyell believes that there formerly existed a centre of permanent eruption in the valley. The Val del Bove is altogether sterile; but the mountain at the same level is, on other sides, clothed with trees. The vast mass of the mountain is realised by the fact that, after twelve miles of the ascent from Catania, the summit looks as far off as it did at starting. Moreover, Mount Vesuvius might be almost hidden away in the Val del Bove.

A remarkable feature of Etna is the large number of minor craters which are scattered over its sides. They look [small in comparison with the great mass of the mountain, but in reality some of them are of large dimensions. Monte Minardo, near Bronte, the largest of these minor cones, is still 750 feet high, although its base has been raised by modern lava-streams which have flowed around it. There are 80 of the more conspicuous of these minor cones, but Von Walters-hausen has mapped no less than 200 within a ten mile radius from the great crater, while neglecting many monticules of ashes. As to the statement made by Reclus to the effect that there are 700 minor cones, and by Jukes, that the number is 600, it is to be supposed that they include not only the most insignificant monticules and heaps of cinders, but also the *bocche* and *boccarelle* from which at any time lava or fire has issued. If these be included, no doubt these numbers are not exaggerations.

The only important minor cone which has been produced during the historical period, is the double mountain known as Monti Rossi, from the red colour of the cinders which compose it. This was raised from the plain of Nicolosi during the eruption of 1669; it is 450 feet in height, and two miles in circumference at the base. In a line between the Monti Rossi and the great crater, thirty-three minor cones may be counted. Hamilton counted forty-four, looking down from the summit towards Catania, and Captain Smyth was able to discern fifty at once from an elevated position on the mountain. Many of these parasitic cones are covered with vegetation, as the names Monte Faggi, Monte Ilice, Monte Zappini, indicate. The names have not been happily chosen; thus there are several cones in different parts of the mountain called by the same name—Monte Arso, Monte Nero, Monte Rosso, Monte Frumento, are the most common of the duplicates. Moreover, the names have from time to time been altered, and it thus sometimes becomes difficult to trace a cone which has been alluded to under a former name, or by an author who wrote before the name was changed. In addition to the minor cones from which lava once proceeded, there are numerous smaller vents for the subterranean fires called *Bocche*, or if very small, *Boccarelle, del Fuoco*. In the eruption of 1669, thirteen mouths opened in the course of a few days; and in the eruption of 1809, twenty new mouths opened one after the other in a line about six miles long. Two new craters were formed in the Val del Bove in 1852, and seven craters in 1865. The outbursts of lava from lateral cones are no doubt due to the fact that the pressure of lava in the great crater, which is nearly 1000 feet in depth, becomes so great that the lava is forced out at some lower point of less resistance. The most northerly of the minor cones is Monte di Mojo, from whence issued the lava of 396 B.C., it is $11\frac{1}{2}$ miles from the crater; the most southerly cone is Monte Ste Sofia, 16 miles from the crater. Nearly all the minor cones are within 10 miles of the crater, and the majority are collected between south-east, and west, that is, in an angular space of 135°, starting

CROTTO DELLE PALOMBE NEAR NICOLOSI

midway between east and south, (45° south of due east) to due west, (90° west of due south). Lyell speaks of the minor cones " as the most grand and original feature in the physiognomy of Etna."

A number of caverns are met with in various parts of Etna ; Boccacio speaks of the Cavern of Thalia, and several early writers allude to the Grotto delle Palombe near Nicolosi. The latter is situated in front of Monte Fusara, and the entrance to it is evidently the crater of an extinct monticule. It descends for 78 feet, and at the bottom a cavern is entered by a long shaft ; this leads to a second cavern, which abruptly descends, and appears to be continued into the heart of the neighbouring Monti Rossi. Brydone says that people have lost their senses in these caverns, " imagining that they saw devils, and the spirits of the damned ; for it is still very generally believed that Etna is the mouth of Hell." Many of the caverns near the upper part of the mountain are used for storing snow, and sometimes as places of shelter for shepherds. We have already seen to what extent Lucretius attributed the eruptions to air pent up within the interior caverns of the mountains.

The surface of the mountain has been divided into three zones or regions—the *Piedimontana* or *Coltivata ;* the *Selvosa* or *Nemorosa ;* and the *Deserta* or *Discoperta.* Sometimes the name of *Regione del Fuoco* is given to the central cone and crater. As regards temperature, the zones correspond more or less to the Torrid, Temperate, and Frigid. The lowest of these, the *Cultivated Region*, yields in abundance all the ordinary Sicilian products. The soil, which consists of decomposed lava, is extremely fertile, although of course large tracts of land are covered by recent lavas, or by those which decompose slowly. In this region the vine flourishes, and abundance of corn, olives, pistachio nuts, oranges, lemons, figs, and other fruit trees.

The breadth of this region varies; it terminates at an approximate height of 2000 feet. A circle drawn with a radius of 10 miles from the crater, roughly defines the limit. The elevation of this on the north is 2,310 feet near Randazzo; on the south, 2,145 feet near Nicolosi; on the east, 600 feet near Mascali; and on the west, 1,145 feet near Bronte. The breadth of this cultivated zone is about 2 miles on the north, east, and west, and 9 or 10 on the south, if we take for the base of the mountain the limits proposed above.

The *Woody Region* commences where the cultivated region ends, and extends as a belt of varying width to an approximate height of 6,300 feet. It is terminated above by a circle having a radius of nearly $1\frac{1}{2}$ miles from the crater. There are fourteen separate forests in this region: some abounding with the oak, beech, pine, and poplar, others with the chestnut, ilex, and cork tree.

The celebrated *Castagno di Cento Cavalli*, one of the largest and oldest trees in the world, is in the Forest of Carpinetto, on the East side of the mountain, five miles above Giarre. This tree has the appearance of five separate trunks united into one, but Ferrara declares that by digging a very short distance below the surface he found one single stem. The public road now passes through the much-decayed trunk. Captain Smyth measured the circumference a few feet from the ground, and found it to be 163 feet, which would give it a diameter of more than 50 feet. The tree derives its name from the story that one of the Queens of Arragon took shelter in its trunk with a suite of 100 horsemen. Near this patriarch are several large chestnuts, which, without a shadow of doubt, are single trees; one of these is 18 feet in diameter, and a second 15 feet, while the *Castagno della Galea*, higher up on the mountain, is 25 feet in diameter, and probably more than 1000 years old. The breadth of the Regione Selvosa varies considerably, as may be seen by reference to the accompanying map; in the direction of the Val del Bove it is very narrow, while elsewhere it frequently has a breadth of from 6 to 8 miles.

The Desert Region is embraced between the limit of 6,300 feet and the summit. It occupies about 10 square miles, and consists of a dreary waste of black sand, scoriœ, ashes, and masses of ejected lava. In winter it remains permanently covered with snow, and even in the height of summer snow may be found in certain rifts.

Botanists have divided the surface of Etna into seven regions. The first extends from the level of the sea to 100 feet above it, and in it flourishes the palm, banana, Indian fig or prickly pear, sugar-cane, mimosa, and acacia. It must be remembered, however, that it is only on the east side of the mountain that the level within the base sinks to 100 feet above the sea; and, moreover, that the palm, banana, and sugar-cane, are comparative rarities in this part of Sicily. Prickly pears and vines are the most abundant products of the lower slopes of the eastern side of Etna. The second, or hilly region, reaches from 100 to 2000 feet above the sea, and therefore constitutes, with the preceding, the *Regione Collivata* of our former division. In it are found cotton, maize, orange, lemon, shaddock, and the ordinary Sicilian produce. The culture of the vine ceases near its upward limit. The third, or woody region, reaches from 2000 to 4000 feet, and the principal trees within it are the cork, oak, maple, and chestnut. The fourth region extends from 4000 to 6000 feet, and contains the beech, Scotch fir, birch, dock, plaintain, and sand-worth. The fifth, or sub-Alpine region, extending from 6000 to 7000 feet, contains the barberry, soapwort, toad-flax, and juniper. In the sixth region, 7,500 to 9000 feet, are found soapwort, sorrel, and groundsel ; while the last narrow zone, 9000 to 9,200 feet, contains a few lichens, the commonest of which is the *Stereocaulon Paschale.* The flora of Etna comprises 477 species, only 40 of which are found between 7000 feet and the summit, while in the last 2000 feet only five phanerogamous species are found, viz., Anthemis Etnensis, Senecio Etnensis, Robertsia taraxacoides, (which are peculiar to Etna), Tanacetum vulgare, and Astragulus Siculus. Common ferns, such as the *pteris aquilina,* are found in abundance beneath the trees in the Regione Selvosa.

This division has been advocated by Presl in his *Flora Sicula.*[1] He names the different regions beginning from below : *Regio Subtropica, Regio Collina, Regio Sylvatica inferior, Regio fagi Sylvestris.* These four are common to all Sicily. The three remaining regions, *Regio Subalpina, Regio Alpina,* and *Regio Lichenum,* together extending from 6000 to 9,200 feet, belong to Etna alone.

At the conclusion of the first volume of Recupero's *Storia Naturale et Generale dell' Etna* we find a somewhat different botanical division proposed by Signor

---

[1] "Flora sicula : exhibens Plantas vasculosas in Sicilia aut sponte crescentes aut frequentissime cultas, secundum systema naturale digestas." Auctore G. B. Presl. Pragæ, 1824.

Rafinesque-Schmaltz.[1]   He makes his divisions in the
following manner :—

1. Florula Piedemontana.
2. Florula Nemorosa.
3. Florula excelsa o della Regione Discoperta.
4. Florula Arenosa o della Regione delle Scorie.

In the latter region, (to which he assigns no limit as to
height), he found Potentilla Argentea, Rumex Scutatus,
Tanacetum Vulgare, Anthemis Montana, Jacobæa Chry-
santhemifolia, Seriola Uniflora, and Phalaris Alpina.

As regards the animal life on Etna, of course it is the
same as that of the eastern sea-board of Sicily, except in
the higher regions, where it becomes more sparse.   The
only living creatures in the upper regions are ants, a
little lower down Spallanzani found a few partridges,
jays, thrushes, ravens, and kites.

Brydone says of the three regions : " Besides the
corn, the wine, the oil, the silk, the spice, and delicious
fruits of its lower region ; the beautiful forests, the flocks,
the game, the tar, the cork, the honey of its second ; the
snow and ice of its third ; it affords from its caverns a
variety of minerals and other productions—cinnabar,
mercury, sulphur, alum, nitre, and vitriol ; so that this
wonderful mountain at the same time produces every
necessary, and every luxury of life."

---

[1] Chloris Aetnensis : o le quattro Florule dell' Etna, opusculo
del Sig. C. S. Rafinesque-Schmaltz, Palermo.   Dicembre, 1813.

CHAPTER III.

ASCENT OF THE MOUNTAIN.

The most suitable time for ascending Etna.—The ascent
commenced.—Nicolosi.—Etna mules.—Night journey through
the upper Regions of the mountain.—Brilliancy of the Stars.—
Proposed Observatory on Etna.—The Casa Inglesi.—Summit of
the Great Crater.—Sunrise from the summit.—The Crater.—
Descent from the Mountain.—Effects of Refraction.—Fatigue of
the Ascent.

THE ascent of Mount Etna has been described many times
during the last eighteen centuries, from Strabo in the
second century to Dr. Baltzer in 1875. One of the most
interesting accounts is certainly that of Brydone, and in
this century perhaps that of Mr. Gladstone. Of course
the interest of the expedition is greatly increased if it can
be combined with that spice of danger which is afforded
by the fact of the mountain being in a state of eruption
at the time.

The best period for making the ascent is between May
and September, after the melting of the winter snows,
and before the autumnal rains. In winter snow frequently extends from the summit downwards for nine or

ten miles; the paths are obliterated, and the guides refuse
to accompany travellers. Even so late in the spring as
May 29th Brydone had to traverse seven miles of snow
before reaching the summit. Moreover, violent storms
often rage in the upper regions of the mountain, and the
wind acquires a force which it is difficult to withstand,
and is at the same time piercingly cold. Sir William
Hamilton, in relating his ascent on the night of June
26th, 1769, remarks that, if they had not kindled a fire
at the halting place, and put on much warm clothing,
they would "surely have perished with the cold." At
the same time the wind was so violent that they had
several times to throw themselves on their faces to
avoid being overthrown. Yet the guides said that the
wind was not unusally violent. Some writers, well used
to Alpine climbing, have asserted that the cold on Etna
was more severe than anything they have ever experienced in the Alps.

The writer of this memoir made the ascent of the
mountain in August 1877, accompanied by a courier and
a guide. We took with us two mules; some thick rugs;
provisions consisting of bread, meat, wine, coffee, and
brandy; wooden staves for making the ascent of the
cone; a geological hammer; a bag for specimens; and a
few other requisites. It has to be remembered that
absolutely nothing is to be met with at the Casa Inglesi,
where the halt is made for the night; even firewood has

to be taken, a fire being most necessary in those elevated regions even during a midsummer's night. For some time previous to our ascent the weather had been uniformly bright and fine, and there had been no rain for more than three months. The mean temperature in the shade at Catania, and generally along the eastern sea-base of the mountain, was 82° F.

As we desired to see the sunrise from the summit of the mountain, we determined to ascend during the cool of the evening, resting for an hour or two before sunrise at the Casa Inglesi at the foot of the cone. Accordingly we left Catania soon after midday, and drove to Nicolosi, twelve miles distant, and 2,288 feet above the sea. The road for some distance passed through a very fertile district; on either side there were corn fields and vineyards, and gardens of orange and lemon trees, figs and almonds, growing luxuriantly in the decomposed lava. About half way between Catania and Nicolosi stands the village of Gravina, and a mile beyond it Mascalucia, a small town containing nearly 4000 inhabitants. Near this is the ruined church of St. Antonio, founded in 1300. Nine miles from Catania the village of Torre di Grifo is passed, and the road then enters a nearly barren district covered with the lava and scoriæ of 1527. The only prominent form of vegetation is a peculiar tall broom—*Genista Etnensis*—which here flourishes. We are now entering the region of minor cones; the vine-clad cone of Monpilieri is visible on the left, and just above it Monti Rossi, 3,110 feet above the sea; to the right of the latter we see Monte San Nicola, Serrapizzuta, and Monte Arso. We reach Nicolosi at half-past four; for although the distance is short, the road is very rugged and steep.

Nicolosi has a population of less than 3,000; it consists of a long street, bordered by one-storied cottages of lava. In the church the priests were preparing for a *festa* in honour of S. Anthony of Padua. They politely took us into the sacristy, and exhibited with much pride some graven images of rather coarse workmanship, which were covered with gilding and bright coloured paint. Near Nicolosi stands the convent of S. Nicola dell' Arena, once inhabited by Benedictine monks, who however were compelled to abandon it in consequence of the destruction produced by successive shocks of earthquake. Nicolosi itself has been more than once shaken to the ground. We dined pretty comfortably, thanks to the courier who acted as cook, in the one public room of the one primitive inn of the town; starting for the Casa Inglesi at 6 o'clock. The good people of the inn surrounded us at our departure and with much warmth wished us a safe and successful journey.

For a short distance above Nicolosi, stunted vines are seen growing in black cinders, but these soon give place to a large tract covered with lava and ashes, with here

and there patches of broom. There was no visible path,
but the mules seemed to know the way perfectly, and
they continued to ascend with the same easy even pace
without any guidance, even after the sun had disappeared
behind the western flank of the mountain. In fact, you
trust yourself absolutely to your mule, which picks his
way over the roughest ground, and rarely stumbles or
changes his even step. I found it quite easy to write
notes while ascending, and even to use a pocket spectroscope at the time of the setting sun. Subsequently we
saw a man extended at full length, and fast asleep upon
a mule, which was leisurely plodding along the highway.
The same confidence must not however be extended to the
donkeys of Etna, as I found to my cost a few days later at
Taormina. Here the only animal to be procured to
carry me down to the sea-shore, 800 feet below, was a
donkey. It was during the hottest part of the day, and
it was necessary to carry an umbrella in one hand, and
comfortable to wear a kind of turban of many folds of
thin muslin round one's cap. The donkey after carefully
selecting the roughest and most precipitous part of the
road, promptly fell down, leaving me extended at full
length on the road, with the open umbrella a few yards
off. At the same time the turban came unfolded, and
stretched itself for many a foot upon the ground. Altogether it was a most comical sight, and it reminded me
forcibly, and at the instant, of a picture which I once

saw over the altar of a church in Pisa, and which represented S. Thomas Aquinas discomfiting Plato, Aristotle,
and Averröes. The latter was completely overthrown,
and in the most literal sense, for he was grovelling in
the dust at the feet of S. Thomas, while his disarranged
turban had fallen from him.

The district of lava and ashes above Nicolosi is succeeded by forests of small trees, and we are now fairly
within the *Regione Selvosa*. At half-past 8 o'clock the
temperature was 66°, at Nicolosi at 4 o'clock it was 80°.
About 9 o'clock we arrived at the Casa del Bosco, (4,216
feet), a small house in which several men in charge of the
forest live. Here we rested till 10 o'clock, and then after
I had put on a great-coat and a second waistcoat, we
started for the higher regions. At this time the air was
extraordinarily still, the flame of a candle placed near the
open door of the house did not flicker. The ascent from
this point carried us through forests of pollard oaks, in
which it was quite impossible to see either a path or any
obstacles which might lie in one's way. The guide
carried a lantern, and the mules seemed well accustomed
to the route. At about 6,300 feet we entered the
Regione Deserta, a lifeless waste of black sand, ashes,
and lava; the ascent became more steep, and the air
was bitterly cold. There was no moon, but the stars
shone with an extraordinary brilliancy, and sparkled like
particles of white-hot steel. I had never before seen

the heavens studded with such myriads of stars. The milky-way shone like a path of fire, and meteors flashed across the sky in such numbers that I soon gave up any attempt to count them. The vault of heaven seemed to be much nearer than when seen from the earth, and more flat, as if only a short distance above our heads, and some of the brighter stars appeared to be hanging down from the sky. Brydone, in speaking of his impressions under similar circumstances says:

"The sky was clear, and the immense vault of heaven appeared in awful majesty and splendour. We found ourselves more struck with veneration than below, and at first were at a loss to know the cause, till we observed with astonishment that the number of the stars seemed to be infinitely increased, and the light of each of them appeared brighter than usual. The whiteness of the milky-way was like a pure flame that shot across the heavens, and with the naked eye we could observe clusters of stars that were invisible in the regions below. We did not at first attend to the cause, nor recollect that we had now passed through ten or twelve thousand feet of gross vapour, that blunts and confuses every ray before it reaches the surface of the earth. We were amazed at the distinctness of vision, and exclaimed together, ‘What a glorious situation for an observatory! had Empedocles had the eyes of Galileo, what discoveries must he not

E

have made!' We regretted that Jupiter was not
visible, as I am persuaded we might have discovered
some of his satellites with the naked eye, or at least
with a small glass which I had in my pocket."

Brydone wrote a hundred years ago, but his idea
of erecting an observatory on Mount Etna was only
revived last year, when Prof. Tacchini the Astronomer
Royal at Palermo, communicated a paper to the
Accademia Gioenia, entitled " *Della Convenienza ed
utilita di erigere sull' Etna una Stazione Astronomico-
Meteorologico.*" Tacchini mentions the extraordinary
blueness of the sky as seen from Etna, and the appearance of the sun, which is " whiter and more tranquil "
than when seen from below. Moreover, the spectroscopic
lines are defined with wonderful distinctness. In the
evening at 10 o'clock, Sirius appeared to rival Venus,
the peculiarities of the ring of Saturn were seen far
better than at Palermo ; and Venus emitted a light
sufficiently powerful to cast shadows ; it also scintillated.
When the chromosphere of the sun was examined the
next morning by the spectroscope, the inversion of the
magnesium line, and of the line 1474 was immediately
apparent, although it was impossible to obtain this effect
at Palermo. Tacchini proposes that an observatory
should be established at the Casa Inglesi, in connection
with the University of Catania, and that it be provided
with a good six-inch refracting telescope, and with

meteorological instruments. In this observatory, constant observations should be made from the beginning of June to the end of September, and the telescope should then be transported to Catania, where a duplicate mounting might be provided for it, and observations continued for the rest of the year. There seems to be every probability that this scheme will be carried out in the course of next year.

During this digression we have been toiling along the slopes of the *Regione Deserta* and looking at the sky; at length we reach the *Piano del Lago* or Plain of the Lake, so called because a lake produced by the melting of the snows existed here till 1607, when it was filled up by lava. The air is now excessively cold, and a sharp wind is blowing. Progress is very slow, the soil consists of loose ashes, and the mules frequently stop; the guide assures us that the Casa Inglesi is quite near, but the stoppages become so frequent that it seems a long way off; at length we dismount, and drag the mules after us, and after a toilsome walk the small lava-built house, called the Casa Inglesi, is reached (1.30 a.m., temperature 40° F.) It stands at a height of 9,652 feet above the sea, near the base of the cone of the great crater, and it takes its name from the fact that it was erected by the English officers stationed in Sicily in 1811. It has suffered severely from time to time from the pressure of snow and from earthquakes, but it was thoroughly

repaired in 1862, on the occasion of the visit of Prince
Humbert, and is now in tolerable preservation. It
consists of three rooms, containing a few deal chairs,
a table, and several shelves like the berths of ships
furnished with plain straw mattresses; there is also
a rough fireplace. We had no sooner reached this
house, very weary and so cold that we could scarcely
move, than it was discovered that the courier had
omitted to get the key from Nicolosi, and there seemed
a prospect of spending the hours till dawn in the open
air. Fortunately we had with us a chisel and a
geological hammer, and by the aid of these we forced
open the shutter serving as a window, and crept into
the house; ten minutes later a large wood fire was
blazing up the chimney, our eatables were unpacked,
some hot coffee was made, and we were supremely
comfortable.

At 3 a.m. we left the Casa Inglesi for the summit
of the great crater, 1,200 feet above us, in order to be
in time to witness the sunrise. Our road lay for a
short distance over the upper portion of the Piano del
Lago, and the walking was difficult. The brighter stars
had disappeared, and it was much darker than it had
been some hours before. The guide led the way with
a lantern. The ascent of the cone was a very stiff
piece of work; it consists of loose ashes and blocks of
lava, and slopes at an angle of "45° or more" according

THE CASA INGLESE AND CONE OF ETNA

to one writer, and of 33° according to another; probably
the slope varies on different sides of the cone: we do not
think that the slope much exceeds 33° anywhere on the
side of the cone which we ascended. Fortunately there
was no strong wind, and we did not suffer from the
sickness of which travellers constantly complain in the
rarefied air of the summit. We reached the highest
point at 4.30 a.m., and found a temperature of 47° F.

When Sir William Hamilton ascended towards the
end of June the temperature at the base of the mountain
was 84° F., and at the summit 56° F. When Brydone
left Catania on May 26th, 1770, the temperature was
76° F., Bar. 29 in. 8½ lines; at Nicolosi at midday on
the 27th it was 73° F., Bar. 27 in. 1½ lines; at the
Spelonca del Capriole (6,200 feet), 61° F., Bar. 26 in.
5½ lines; at the foot of the crater, temp. 33° F., Bar.
20 in. 4½ lines, and at the summit of the crater just
before sunrise, temp. 27° F., Bar. 19 in. 4 lines.

On reaching the summit we noticed that a quantity
of steam and sulphurous acid gas issued from the ground
under our feet, and in some places the cinders were so
hot that it was necessary to choose a cool place to sit
down upon. A thermometer inserted just beneath the
soil from which steam issued registered 182° F. For a
short time we anxiously awaited the rising of the sun.
Nearly all the stars had faded away; the vault of
heaven was a pale blue, becoming a darker and darker

grey towards the west, where it appeared to be nearly black. Just before sunrise the sky had the appearance of an enormous arched spectrum, extremely extended at the blue end. Above the place where the sun would presently appear there was a brilliant red, shading off in the direction of the zenith to orange and yellow ; this was succeeded by pale green, then a long stretch of pale blue, darker blue, dark grey, ending opposite the rising sun with black. This effect was quite distinct, it lasted some minutes, and was very remarkable. This was succeeded by the usual rayed appearance of the rising sun, and at ten minutes to 5 o'clock the upper limb of the sun was seen above the mountains of Calabria. Examined by the spectroscope the Fraunhofer lines were extremely distinct, particularly two lines near the red end of the spectrum.

The top of the mountain was now illuminated, while all below was in comparative darkness, and a light mist floated over the lower regions. We were so fortunate as to witness a phenomenon which is not always visible, viz., the projection of the triangular shadow of the mountain across the island, a hundred miles away. The shadow appeared vertically suspended in space at or beyond Palermo, and resting on a slightly misty atmosphere ; it gradually sank until it reached the surface of the island, and as the sun rose it approached nearer and nearer to the base of the mountain. In a short time the flood of light destroyed the first effects of light and shadow. The mountains of Calabria and the west coast of Italy appeared very close, and Stromboli and the Lipari Islands almost under our feet; the east coast of Sicily could be traced until it ended at Cape Passaro and turned to the west, forming the southern boundary of the island, while to the west distant mountains appeared. No one would have the hardihood to attempt to describe the various impressions which rapidly float through the mind during the contemplation of sunrise from the summit of Etna. Brydone, who is by no means inclined to be rapturous or ecstatic in regard to the many wonderful sights he saw in the course of his tour, calls this "the most wonderful and most sublime sight in nature." "Here," he adds, "description must ever fall short, for no imagination has dared to form an idea of so glorious and so magnificent a scene. Neither is there on the surface of this globe any one point that unites so many awful and sublime objects. The immense elevation from the surface of the earth, drawn as it were to a single point, without any neighbouring mountains for the senses and imagination to rest upon, and recover from their astonishment in their way down to the world. This point or pinnacle, raised on the brink of a bottomless gulph, as old as the world, often discharging rivers of fire and throwing out burning rocks with a noise that shakes the whole island.

Add to this, the unbounded extent of the prospect, comprehending the greatest diversity and the most beautiful scenery in nature, with the rising sun advancing in the east to illuminate the scene."

When the sun had risen we had time to examine the crater, a vast abyss nearly 1000 feet in depth, and with very precipitous sides. Its dimensions vary, but it is now between two and three miles in circumference. Sometimes it is nearly full of lava, at other times it appears to be bottomless. At the present time it is like an inverted cone; its sides are covered with incrustations of sulphur and ammonia salts, and jets of steam perpetually issue from crevices. Near the summit we found a deposit, several inches in thickness, of a white substance, apparently lava decomposed by the hot issuing gases. Hydrochloric acid is said to frequently issue from the crater; the gases that were most abundant appeared to be sulphurous acid and steam. The interior of the crater appeared to be very similar to that of the Solfatara near Puzzuoli. During the descent from the cone we collected various specimens of ash and cinder, some red, others black and very vesicular, others crystalline, some pale pink. The steep slope of the cone was well shown by the fact that, although the surface is either extremely rugged owing to the accumulation of masses of lava, or soft and yielding on account of the depth of cinders, a large mass of lava set rolling at the top rushes down with increasing velocity until it bounds off to the level plain below.

The great cone is formed by the accumulation of sand, scoriæ, and masses of rock ejected from the crater; it is oval in form, and has varied both in shape and size in the course of centuries. When we saw it, it was not full of smoke or steam; but it was possible to see to the bottom of it, in spite of small jets of steam which issued from the sides. It presented the appearance of a profound funnel-shaped abyss; the sides of which were covered with an efflorescence of a red or yellow, and sometimes nearly white, colour. The crater presented the same appearance when it was seen by Captain Smyth in 1814, but he was so fortunate as to witness it in a less quiescent state. "While making these observations," he writes, "on a sudden the ground trembled under our feet, a harsh rumbling with sonorous thunder was heard, and volumes of heavy smoke rolled over the side of the crater, while a lighter one ascended vertically, with the electric fluid escaping from it in frequent flashes in every direction. . . . . During some time the ground shook so violently that we apprehended the whole cone would tumble into the burning gulf (as it actually had done several times before) and destroy us in the horrible consequences; however, in less than a couple of hours all was again clear above and quiet within." When Mr. Gladstone ascended in 1838, the volcano was in a

slight state of eruption: "The great features of this
action," he writes, "are the sharp and loud claps,
which perceptibly shook from time to time the ground of
the mountain under our feet; the sheet of flame which
leapt up with a sudden momentary blast, and soon disappeared in smoke; then the shower of red-hot stones
and lava. At this time, as we found on our way down,
lava masses of 150 or 200 pound weight were being
thrown a distance of probably a mile and a half; smaller
ones we found even more remote. These showers were
most copious, and often came in the most rapid succession.
Even while we were ascending the exterior of the cone,
we saw them alighting on its slope, and sometimes
bounding down with immense rapidity within, perhaps,
some thirty or forty yards of our rickety footing on
the mountain side. They dispersed like the sparks
of a rocket; they lay beneath the moon, over the
mountain, thicker than ever the stars in heaven; the
larger ones ascended as it were with deliberation, and
descended, first with speed and then with fury. Now
they passed even over our heads, and we could pick up
some newly fallen, and almost intolerably hot. Lastly,
there was the black grey column, which seemed smoke,
and was really ash, and which was shot from time to
time out of the very bowels of the crater, far above its
edge, in regular unbroken form."

At the Casa Inglesi we remounted the mules, and

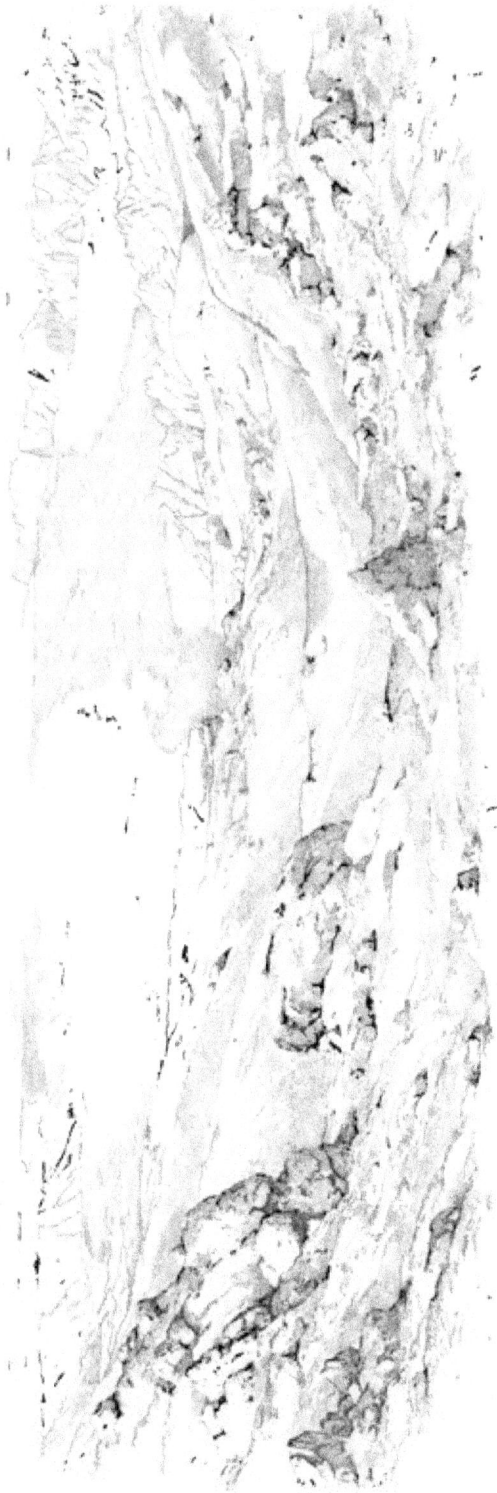

VIEW OF THE VAL DEL BOVE.

made a slight detour to the east in order to look down into the Val del Bove, which is here seen as a gigantic valley, bounded on the north by the precipitous cliffs of the Serra delle Concazze, and on the South by the Serra del Solfizio. It is believed by Lyell and others that in the Balzo di Trifoglietto, at which point the precipices are most profound and abrupt, there was a second permanent crater of eruption. The Torre del Filosofo, a ruined tower, traditionally the observatory of Empedocles, stands near the Casa Inglesi. Not far from this a great deposit of ice was found in 1828. It was preserved from melting by a layer of ashes and sand, which had covered it, soon after its first existence, as a glacier: a stream of lava subsequenly flowed over the ashes, and completely protected the ice; the non-conducting power of the ashes prevented the lava from melting the ice. The snow which falls on the mountain is stowed away in caves, and used by the Sicilians during summer. A ship load is also sent to Malta, and the Archbishop of Catania derives a good deal of his income from the sale of Etna snow.

During our descent from the mountain we were much struck by the apparent nearness of the minor cones beneath us, and of the villages at the base of the mountain. They seemed to be painted on a vertical wall in front of us, and although from ten to fifteen miles distant they appeared to be almost within a stone's throw. This

curious effect, which has often been observed before, is
due to refraction. At the summit of Etna we have left
one-third of the atmosphere beneath us, and the air is
now pressing upon the surface of the earth with a weight
of ten pounds on the square inch, instead of the usual
fifteen pounds experienced at the level of the sea. In
looking towards the base of the mountain we are consequently looking from a rarer to a denser medium ; and
it is a law of optics, that when light passes from a
denser to a rarer medium it is refracted away from the
perpendicular, and thus the object, from which it
emanates, appears raised, and nearer to us than it really
is. The objects around Etna appear near to us and
raised vertically from the horizon for the same reason
that a stick plunged in water appears bent.

We reached Nicolosi again about noon, having left it
eighteen hours before. The ascent of the mountain,
although it does not involve much hard walking, is
somewhat trying on account of the extremes of temperature which have to be endured. In the course of the
morning of our descent we had experienced a difference
equal to more than 40° F. As to the ascent, you are
moving upwards nearly all night ; you have six hours
of riding on a mule, some of it in a bitterly cold atmosphere ; you get very much heated by the final steep climb
of 1100 feet, and you find at the summit a piercing
wind ; of course there is no shelter, and you sit down to

wait for sunrise on cinders which are gently giving off steam and sulphurous acid ; the former condenses to water as soon as it meets the cold air, and you find your great coat, or the rug on which you have sat down, speedily saturated with moisture.

CHAPTER IV.

TOWNS SITUATED ON THE MOUNTAIN.

Paterno.—Ste. Maria di Licodia.—The site of the ancient town of Aetna.—Biancavilla.—Aderno.—Sicilian Inns.—Adranum.—Bronte.—Randazzo.—Mascali.—Giarre.—Aci Reale.—Its position.—The Scogli de'Ciclopi.—Catania, its early history, and present condition.

We have before alluded to the fact that Etna is far more thickly populated than any other part of Sicily or Italy; in fact, more so than almost any equal area in the world, of course excepting large cities and their neighbourhood. This is due to the wonderful fertility of the soil, the salubrity of the climate, and, on the eastern base, to the proximity of a sea-coast indented with excellent harbours. The habitable zone of Etna is restricted to the *Regione Coltivata*, nevertheless some of the towns on the north and west have a considerable elevation; thus Bronte is 2,782 feet above the sea, and Randazzo 2,718. All the principal towns are situated on the base road of the mountain, which was indeed constructed in order to connect them. Out of the sixty-four towns and villages on the mountain, the following are the most important: Catania, Aci Reale, Paterno, Aderno, Bronte, Randazzo, Aci S. Antonio, Biancavilla, Calatabiano, Giarre, Francavilla, Linguagrossa, Licodia, Mascali, Misterbianco, Nicolosi, Pedara, Piedemonte, Trecastagne, and Tremestieri.

On our return from the summit, we rested for awhile at Nicolosi, and in the cool of the evening started to make a *giro* of the mountain by way of the base road. Descending by the Nicolosi road as far as Mascalucia, we branched off to the west, and made for Paterno, passing near the town of Belpasso, which was destroyed by the earthquake of 1669, and subsequently erected on a new site. It still contains more than 7,000 inhabitants, although the district is extremely unhealthy.

Paterno, the second largest town on the flanks of Etna after Catania and Aci Reale, stands in the very heart of the Regione Coltivata, and possesses more than 16,000 inhabitants. According to Cluverius, it is the site of the city of Hybla Major ("Υβλα Μεγάλη), a Sikelian city which was unsuccessfully attacked by the Athenians soon after they first landed in Sicily. During the second Punic War, the inhabitants went over to the Carthagenians, but the city was speedily recovered by the Romans. Pliny, Cicero, and Pausanias allude to it, but its later history has not come down to us. An altar was lately found in Paterno dedicated to *Veneri Victrici Hyblensi.* Several towns in Sicily were called Hybla, probably—according to Pausanias—in honour of a local deity. Paterno was founded by Roger I. in 1073 : it was once a feudal city of some importance, and possessed a cathedral and castle, and several large monasteries. Athough much fallen to decay, it still possesses a good deal of vitality, and the population is on the increase.

On leaving Paterno the road turns to the North-west, and passes through the village of Ste. Maria di Licodia. Here originally stood the Sikelian City of Inessa ("Ινησσα), which, after the death of Hiero I., was peopled by colonists from Katana (then called Αιτνη). The new occupants of the city changed its name from Inessa to Aetna, which it retained. The town later fell into the hands of the Syracusans, and in 462 B.C. the Athenians in vain attempted to take it. During the Athenian expedition both Aetna and Hybla were allies of Syracuse. In 403 B.C. Aetna was taken by Dionysius, who placed in it a body of Campanian mercenaries. Sixty-four years later (B.C. 339) the town was taken by Timoleon. For many succeeding years we find no further mention of it. Cicero speaks of it in his time as an important place, and the centre of a very fertile district ; it is also mentioned by Pliny and Ptolemy, and Strabo says that it was usually the starting point for those who ascended the mountain. Of its later history we know absolutely nothing.

Six miles to the north-west of St. Mariah di Licodia, the road passes through Biancavilla—a town of 13,000 inhabitants, and the centre of a cotton district.

The road continues in the same direction until the town of Aderno is reached; and here we arrived late in the evening, and gained our first experience of a Sicilian inn in an out-of-the-way town. After many enquiries we were directed to the only inn which the place could boast, kept by a doctor. No one appeared at or near the entrance, of course there was no bell or knocker, and we made our way up a dark stone staircase till we arrived at a dimly-lighted passage. A horrible old Sicilian woman now appeared, and showed us with great incivility the only room in the house, which its inmates were willing to place at our disposal. It was a fairly large room, with a stone floor which apparently had not been swept for weeks, and walls that had once been whitewashed; the furniture consisted of three beds placed on tressels, a plain deal table, and some primitive chairs. As to food they had neither bread, meat, wine, eggs, macaroni, fruit, or butter in the house; neither did they offer to procure anything. Even when some eggs had been obtained, and (after an hour's delay) cooked, there was not a single teaspoon to eat them with. The people of the town appear to subsist chiefly on beans and a kind of dried fish. If our courier had not been a very handy fellow and a tolerable cook, we

F

should have been obliged more than once to go to bed supperless. As it was, the best he could do on this occasion was to get some bread, eggs, and wine, and—best of all—some snow, for the heat was intolerable. In a town of the same size—15,657 inhabitants—in England, we should have at least two really comfortable inns ready at any moment to receive and entertain the weary traveller.

Aderno stands on the site, and has preserved the name, of the ancient Sikelian city of Adranum (Αδρανον). According to Diodorus there existed here, from very early times, the temple of a local deity named Adranus. The city was founded by the elder Dionysius in 400 B.C.; it owed its importance to the renown of its temple, which was guarded by a thousand dogs. In 345 B.C. the city fell into the hands of Timoleon, and it was taken by the Romans at the commencement of the first Punic War. After this we cease to hear of it. The modern town was founded by Roger I. in the 12th century. The fine Norman tower—now used as a prison—and the monastery, were both built by King Roger.

After leaving Aderno the base-road ascends, turns nearly due north, and leads us past a number of lava streams, notably those of 1610, 1603, and 1651. A good view of Monte Minardo, and the minor cones in its more immediate neighbourhood, is obtained on the

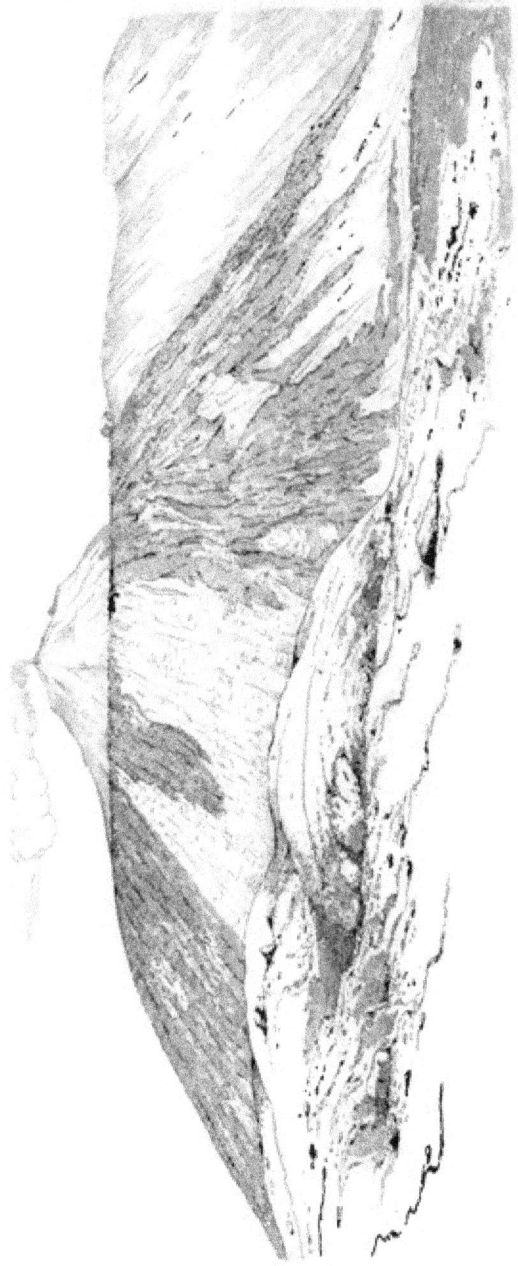

left, while on the right we see the Valley of the Simeto,
and Centorbi high upon the hills.

Nearly due west of the great crater is the town of
Bronte, which is 2,782 feet above the sea, and has a
population of more than 15,000. It is a very primitive
place, and several centuries behind the age ; it reminded
us forcibly, in one or two particulars, of Pompeii : the
streets are narrow and tortuous, and the roadway very
uneven. Awnings are sometimes hung across the street
from side to side to provide shade. The shops are
exactly like those at Pompeii ; and in the main street
we noticed an open-air kitchen, to which the would-be
diner repairs, purchases a plateful of food, and eats
it standing in the public way. The inn was even
worse than that of Aderno, and apparently had never
before received guests. We were offered one miserable
room, without a lock to the door, and unprovided with
either table or chair. Of course the bare idea of offering
to procure, or furnish, or cook, any kind of food was
too monstrous to be entertained for a moment. With
difficulty the courier obtained some eggs, macaroni, and
fruit, on which we dined in a small barn attached to a
wine-shop.

At Bronte we are only nine miles from the crater, on
the steepest side of the mountain, and near the Tertiary
sandstone which underlies this portion of the mountain.
A short distance outside the town we saw great beds

of the lava of 1832, piled up fantastically in all sorts
of forms, and excessively rugged and uneven. It is
quite bare of vegetation, and does not appear to have
even commenced to be decomposed.

Bronte gave its name to Lord Nelson, who was
created Duke of Bronte by Ferdinand IV.:—an appro-
priate name for a great warrior (βροντή, thunder). The
Nelson estates are scattered around the town.

On leaving Bronte the road conducted us past several
high hills of sandstone and quartzite near Monte
Rivoglia; then we passed near Maletto, and, leaving
the malarious lake Gurrita on our left, we soon after
arrived at Randazzo. Near Maletto the road reaches
it highest point—3,852 feet.

The town of Randazzo was founded by the Lombards
in the 10th century; during the Middle Ages it appears
to have been a prosperous, populous, place; at present it
possesses more than 8,000 inhabitants. The Emperor
Frederick II. created his son Duke of Randazzo, and
added to the name of the town, *Etnea.* It contains
several very interesting architectural remains; a church
of the 13th century, a mediæval palace—the Palazzo
Finochiaro,—and a ducal palace now used as a prison.
The houses are for the most part built of lava, and some
of the shops have massive lava counters extending half
across their open front, while the door occupies the
remainder, as at Pompeii. The view from Randazzo is

very fine in every direction; the crater of Etna appears
near, and Monte Spagnuolo—many hours distant—just
outside the town. The town is 2,718 feet above the
sea, just above the Valley of the Alcantara—of which
it commands a fine view, and also of the limestone hills
on the other side.

We were obliged to pass the night in the town, in an
inn scarcely superior to that of Aderno, but distinctly
better than the miserable Albergo Collegio at Bronte.
At least the people were civil, and did their best. The
one room of the inn had a bed in each corner, and a
deal table in the middle. Three of the beds were
occupied by engineers who were surveying in connection
with a new line of railway; the fourth was made over
to the courier. I slept in a small kind of ante-room on
a bed chiefly composed of deal boards placed on tressels.
Here again the courier was invaluable, in fact it would
be simply impossible to make the circuit of Etna without
a courier. He procured some eggs, macaroni, fruit,
snow, tomatoes, and even meat, and cooked everything
well, without a trace of garlic. He also took care that
the linen was clean, and the general arrangements as
comfortable as they could be under the circumstances.
Let us also admit that neither at Aderno, Bronte, nor
Randazzo were we troubled with musquitoes or any
worse species of insect. These, we were assured, would
appear in full force in the following month (September).

Our only inconvenience of this nature arose from swarms of flies. The inns of these out-of-the-way towns probably receive scarcely a dozen travellers in the year, and these are Sicilians, who are not used to better accommodation. Evidently a *forestiare* is quite a novelty : the people of these small towns used to look at us with great curiosity, and crowded round the carriage when we started. At Bronte we had a good example of this curiosity : owing to the hardness of the lava of 1832 the head had come off the handle of our hammer, and we went into a carpenter's shop to have it put on again. Presently we noticed that eleven people, including a priest, were looking on, apparently with intense and absorbing interest.

From Randazzo the base-road descends, until at Giarre it is near the sea-level. This road is one of the most beautiful in Sicily; it is part of the old military route from Messina to Palermo, and it was traversed by Himilco in 396 B.C.; by Timoleon in 344 B.C. ; and by Charles V. in 1534. After leaving Randazzo the valley of the Alcantara becomes visible, while beyond it rise the lofty mountains of the Nebrodes. The road passes near Monte Dolce, and soon reaches Linguaglossa, a small town from whence the craters of 1865 may be reached in about four hours. The rapidly descending road passes through Piedemonte and Mascali, in the heart of an extraordinarily fertile region. Mascali, a

village of 3050 inhabitants, was considered by Cluverius
to be the site of the Greek town of Callipolis, founded
by a colony from Naxos as early as the fifth century,
B.C. A full view of the coast line is obtained from
the Capo di Taormina on the north, to a point below
Riposto on the south. We descended through plantations of nuts, and groves of oranges and lemons, to
gentle slopes covered with vineyards.

From the town of Giarre, (17,965 inhabitants), we
get a view of the Val del Bove, which, however, is
almost always obscured by thin white clouds, while
the summit of the mountain is clear. We noticed,
indeed, every day that the summit, which had been
absolutely clear all the day and night, became covered
with clouds shortly before sunset, while about an
hour later the clouds cleared off, and the mountain
was sharply defined against the sky during the starlit
night. Some of the effects of sunset behind clouds
resting on the summit, while all the rest of the sky was
bright blue, were exceedingly beautiful, and were quite
untranslatable into any known language, save that of
painting, and of music. Perhaps Turner could have
done justice to them.

After leaving Giarre we passed through a good deal
of highly cultivated land belonging to Baron Pennisi,
the largest landholder and richest man in Sicily. He
makes good use of his wealth, and seems to be very

popular among all classes. He possesses three palaces
in Aci Reale, and has done a great deal to beautify
the town. Archæologists will remember him as the
possessor of the finest collection of Sicilian coins in
the world. Many of these have been found on his own
estates, but he never scruples to give large sums of
money for any coin which he covets.

Aci Reale, one of the prettiest towns in Europe,
is situated in the midst of a very fertile region 550
feet above the sea. To the east it faces the Ionian
sea, while on the west towers Etna. The town is
full of wealthy inhabitants, and the houses are large,
lofty, and well built. It contains 24,151 inhabitants,
and possesses celebrated sulphur baths, and one of the
best hotels in Sicily. The wealth of this small town
is well shown by the following fact: Since its foundation in the tenth century, till within a year or two
of the present time, the town had been under the
jurisdiction of the Archbishop of Catania. It happened,
however, a few years ago, on the occasion of a religious
procession in Catania, that the people of Aci considered
that their patron Saint, S. Venera, was slighted. In
fact the image of S. Agata, the patron Saint of the
Catanese—whose veil has so often averted the lava-
streams from the city—was put in all the prominent
parts of the procession, while the image of S. Venera
was comparatively neglected. The people of Aci at once

returned home, and sent a petition to the Pope, praying that they might have a Bishop of their own directly subject to the Holy See, in order that they might no longer be subjected to such slights. The Vatican having duly considered the question consented to raise Aci to the dignity of a Bishopric, and to pay the Bishop a yearly stipend of 10,000 lire, (about £400, but equal to £600 in Sicily), on condition that 200,000 lire were paid at once into the coffers of the Vatican. This was promptly done, and now Monsignore Gerlando Gennardi, Bishop of Aci Reale, may snap his fingers in the face of Monsignore Giuseppe Benedetto Dusmet, a Benedictine of the Congregation of Monte Cassino, and Archbishop of Catania.

Six villages in the neighbourhood of Aci Reale bear the name of Aci : Aci Castello, Aci Sant' Antonio, and so on, but Aci Reale claims to stand upon the very site rendered memorable by the story of Acis and Galatea. The river Acis (now called *Acque Grande*) rises from a bed of lava, and falls into the sea a mile from its source. Aci Reale stands on seven different beds of superposed lava, having layers of earth resulting from decomposed lava between. The Canon Recupero calculated from observation, that a lava requires at least 2000 years to form even a scanty layer of earth, consequently he inferred that the lowest of the lava streams upon which Aci rests must have been formed 14,000 years ago. These views he stated to Brydone a hundred years ago; the latter says, "Recupero tells me he is exceedingly embarrassed by these discoveries in writing the history of the mountain. That Moses hangs like a dead weight upon him, and blunts all his zeal for enquiry; for that really he has not the conscience to make his mountain so young as that prophet makes the world. What do you think of these sentiments from a Roman Catholic Divine? The Bishop, who is strenuously orthodox—for it is an excellent See—has already warned him to be upon his guard, and not to pretend to be a better natural historian than Moses; not to presume to urge anything that may, in the smallest degree, be deemed contradictory to his sacred authority." The Canon Recupero lost his church preferment on the publication of Brydone's book, and the whole body of clergy of Girgenti received a reprimand on account of a capital story which Brydone told of a dinner at which the Bishop presided, during which several of the reverend Canons suffered severely from the effects of English punch, which Brydone had brewed for them. We quite agree with Admiral Smyth when he says, "It is a pity that Mr. Brydone laboured under such a cacoethes, as to sacrifice a friend for the sake of a good story." Of course we now know that Recupero's estimate of the age of Etna was far within the true

ISLAND

J. LEITCH & C⁰ Sc

ASALT OFF TREZZA

limits, but we derive this information from other sources. No true estimate can be obtained from the observation of the decomposition of lavas, for it has been often observed that two lavas will decompose at very different rates.

A little to the north of the village of La Scaletta, at the base of the rocks upon which Aci Reale stands, there are two small caverns in the abrupt face of the basalt, which can only be approached in a boat. They consist of columnar basalt bent very curiously, and capped by amorphous basalt.

A drive of a few miles to the South of Aci Reale brings us to Trezza, a small village built of lava. A short distance from the shore are the celebrated *Scogli di Ciclopi*, or rocks of the Cyclops, said to be those which Polyphemus hurled at Ulysses after his escape from the cave. The rocks, seven in number, form small islets, the largest of which, the Isola d' Aci, is about 3000 feet in circumference, and 150 feet high. It consists of crudely columnar basalt capped by a kind of marl. Near the top of the island there is a cave called the " Grotto of Polyphemus," also a cistern of water. To the south of this island a very picturesque rock rises from the sea. It is 2000 feet in circumference and about 200 feet in height, and consists of columnar basalt in four and eight-sided prisms, but not very regular; a hard calcareous substance is

found in their interstices. Fine crystals of analcime
are sometimes met with in the basalts of the Cyclops
Islands. Lyell considers these basalts "the most ancient
monuments of volcanic action within the region of Etna."

A few miles south of the Isole di Ciclopi are the
bay and city of Catania. We started from the latter
when we commenced our ascent of Etna, and now on
returning to it, we completed the circuit of the mountain
by its base-road of 87 miles.

Katana (Κατάνη) is believed to have been founded
about 730 B.C. by a Greek colony of Naxos, which had
originally come from Chalcis. The city maintained its
independence till the time of Hieron, who expelled the
original inhabitants in 476 B.C., and peopled the city
with Syracusans and inhabitants of the Peloponnesus
to the number of 10,000. At the same time the name
of the city was changed to Aetna (Αἴτνη). In 461 B.C.,
however, the old inhabitants retook their city, and drove
out the newly-settled strangers, who betook themselves
to Inessa, occupied it, and changed its name to *Aetna*.
At a later period the Katanians sided with the Athenians
against the Syracusans. But in 403 B.C. Dionysius of
Syracuse took and plundered the city, sold the inhabitants as slaves, and established in it a body of Campanian
mercenaries. The latter quitted it and retired to Aetna
in 396 B.C., when the city was taken by the Car-
thaginians after a battle off the rocks of the Cyclops.

Katana submitted to the Romans in 263 B.C., during the first Punic War, and it soon became a very populous city. Cicero mentions it as a wealthy city and important seaport. During the Middle Ages it underwent many changes both at the hands of nature and of man; it belonged in succession to the Goths, Saracens, and Normans; and in 1169 was destroyed by an earthquake, during which 15,000 of its inhabitants perished. Again in 1669, and 1693, it was almost destroyed by earthquakes. The present town is comparatively new, many of its more ancient remains are covered with lava, among them the remains of a fine Greco-Roman theatre, in which it is probable that Alcibiades addressed the Catanians in 415 B.C. There are also remains of a Roman amphitheatre, bath, and tombs. Of more modern structures, the cathedral is the first to claim our notice. It was commenced by Roger I. in 1091, but in less than a century was almost entirely destroyed by an earthquake. At one corner of the building you descend through a narrow passage cut in the lava, to a crypt in which some ancient Roman arches are shown, partly filled up with lava. Here also is seen a small stream of very clear water flowing through the lava. The cathedral contains several interesting tombs, and in the chapel of S. Agata, her body is preserved in a silver sarcophagus, which during certain fetes is carried through the town in procession, attended by all the authorities. S. Agata was martyred by the Prætor Quintianus in the reign of Decius, and is the patron saint of the city. Whenever Catania has been in trouble from the approach of lava streams, or from earthquakes, the veil of S. Agata has been used as a charm to avert the evil.

The University of Catania is the most celebrated in Sicily. It was founded in 1445 by Alfonso of Arragon, and has produced several men of eminence. The city also possesses one of the finest monasteries in the world, now converted into schools and barracks. Formerly the monastery of S. Nicola was occupied by 40 monks, all members of noble families; it is sufficiently large to hold 400.

CHAPTER V.

ERUPTIONS OF THE MOUNTAIN.

Their frequency within the historical period.—525 B.C.—477
B.C.—426 B.C.—396 B.C.—140 B.C.—134 B.C.—126 B.C.—122 B.C.—
19 B.C.—43 B.C.—38 B.C.—32 B.C.—40 A.D.—72.—253.—420.—812
—1169.—1181.—1285.—1329.—1333.—1371.—1408.—1444.—
1446.—1447.—Close of the Fifteenth Century.—1536.—1537—
1566.—1579.—1603.—1607.—1610.—1614.—1619.—1633.—1646.
—1651.—1669.—1682.—1688.—1689.—1693.—1694.—1702.—
1723.—1732.—1735.—1744.—1747.—1755.—Flood of 1755.—
1759.—1763.—1766.—1780.—1781.—1787.—1792.—1797.—1798.
—1799.—1800.—1802.—1805.—1808.—1809.—1811.—1819.—
1831.—1832.—1838.—1842.—1843.—1852.—1865.—1874.—
General character of the Eruptions.

A list of all the eruptions of Etna from the earliest
times has been given by several writers, notably by
Ferrara in his *Descrizione dell' Etna*, and by Gemellaro.
The latter places the first eruption in 1226 B.C. in the
time of the Sicani; the second in 1170 B.C.; and of the
third he says, "In 1149 B.C. there was an eruption, and
Hercules in consequence fled from the island." Of
course these dates are worthless. and the statements are

no doubt based upon the assertion of Diodorus, that before the Trojan war the Sicani were driven from the east side of Sicily by the eruptions of the volcano.

1. The first eruption appears to have occurred in the time of Pythagoras; we have no details as to its nature.

2. The second eruption occurred in 477 B.C. It is mentioned by Thucydides, and it must be the eruption to which Pindar and Æschylus allude. The former visited the tomb of Hiero I. of Syracuse in 473 B.C., and the latter was in Sicily in 471 B.C. On the occasion of this eruption, two heroic youths named Anăpias and Amphĭnomus, performed a deed to which Seneca and other writers allude with enthusiasm. While the lava was rapidly overwhelming the city of Katana, they placed their aged parents on their shoulders, and, at the risk of their lives, bore them through the flaming streets, and succeeded in placing them in safety. It was said that the fiery stream of lava parted to make way for them. Statues were raised to the honour of the *Pii Fratres*, and their burial place was long known as the *Campus Piorum*. Even a temple was erected to commemorate the deed.

Lucilius Junior devotes the concluding lines of his poem on Etna to the glory of the brothers : "The flames blushed to touch the filial youths, and wherever they plant their footsteps, they retire. That day is a day of fortune; harmless that land. On their right hand fierce dangers prevail; on their left are burning fires. Athwart the flames they pass in triumph, his brother and he, each safe beneath his filial burden. There the devouring fire flees backward, and checks itself round the twin pair. At length they issue forth unharmed, and bear with them their deities in safety. Songs of poets honour and admire them; them has Pluto placed apart beneath a glorious name, nor can the mean Fates reach the holy youths, but have truly granted them the homes and dominion of the blessed." [1]

3. The third eruption occurred in the year 426 B.C. It is mentioned by Thucydides as having commenced in the sixth year of the Peloponnesian War. It destroyed a portion of the territory of the inhabitants of Katana.

4. An important eruption occurred in the year 396 B.C. It broke out from Monte di Mojo, the most northerly of the minor cones of Etna, and following the course of the river Acesines, (now the Alcantara) entered the sea at the site of the ancient Greek colony of Naxos. Himilco the Carthaginian general, was at this time on his way from Messana to Syracuse, and he was compelled to march his troops round the west side of the mountain in order to avoid the stream of lava.

5. We hear of no further eruption for 256 years, when in the year 140 B.C., in the consulship of C. Lælius

---

[1] Translated by L. E. Upcott, M.A.

Sapiens and Q. Servilius Cœpio, there was an outburst from the volcano which destroyed 40 people.

5. Six years later an eruption occurred according to Orosius and Julius Obsequens, in the consulship of Sergius Fulvius Flaccus, and Quintus Calpurnius Piso. We have no details concerning its nature or extent.

7. The same authorities state that in the year 126 B.C. in the consulship of L. Œmilius Lepidus, and L. Aurelius Orestes, Sicily suffered from a very severe earthquake, and a deluge of fiery matter poured from Etna, over-whelming large tracts of country, and rendering the waters of the adjacent Ionian sea positively hot. It is said that the sea near the island of Lipari boiled, and that the inhabitants ate so large a number of the fishes which were cast, already cooked, upon their shores, that a distemper appeared which destroyed a large number of people.

8. Four years later Katana was nearly destroyed by a new eruption. The roofs of many of the houses were broken in by the weight of hot ashes which fell upon them ; but the lava stream turned aside near the city and flowed into the sea. The lava is believed to have issued from a small crater near Gravina, about 2½ miles from Katana. The city was so much injured by this eruption that the Romans granted the inhabitants an immunity from all taxes for a space of ten years.

9. An eruption, of which we have no details, occurred during the civil war between Cæsar and Pompey.

10. Livy speaks of an eruption and earthquake which took place shortly before the death of Cæsar, which it was believed to portend.

11. In 38 B.C., during the civil war between Octavianus and Sextus Pompeius, a violent eruption occurred on the east side of the mountain, accompanied by fearful noises and outbursts of flame.

12. Six years afterwards an eruption of a less violent character took place.

13. The next eruption of which we hear is that mentioned by Suetonius in his Life of Caligula. The Emperor happened to be at Messina at the time, and he fled from the town through fear of the eruption. This was in 40 A.D.

14. An eruption is said to have occurred in 72, in the second year after the capture of Jerusalem by Titus.

15. Etna was now quiescent for nearly two centuries, but in the year 253, in the reign of the Emperor Decius, a violent eruption lasting nine days occurred. The lava flowed in the direction of Catania, and the inhabitants for the first time tested the efficacy of the veil of S. Agatha, which afterwards stood them in such good stead on more than one occasion. The Saint had been martyred the year before, and when the frightened

inhabitants saw the stream of lava approaching the city, they rushed to the tomb, and removed the veil which covered her body. This was carried to the edge of the descending torrent of lava, and is asserted to have at once arrested its progress.

16. According to Carrera and Photius an eruption occurred in the year 420.

17. We now find no record of any volcanic action for nearly four hundred years. Geoffrey of Viterbo states that an eruption occurred in 812, when Charlemagne was in Messina.

18. After another long interval of more than three centuries and a half, the mountain again entered into eruption. In February, 1169, occurred one of the most disastrous eruptions on record. A violent earthquake, which was felt as far as Reggio, occurred about dawn, and in a few minutes Catania was a heap of ruins. It is estimated that 15,000 persons were buried beneath the ruins. It was the vigil of the feast of S. Agatha, and the Cathedral of Catania was crowded with people, who were all buried beneath the ruins, together with the Bishop and forty-four Benedictine monks. The side of the cone of the great crater towards Taormina fell into the crater. At Messina the sea retired to some distance from the shore, and then suddenly returned, overwhelming a portion of the city, and sweeping away a number of persons who had fled to the shore for

safety. The clear and pure fountain of Arethusa at Syracuse became muddy and brackish; while the fountain of Ajo, near the village of Saraceni, ceased to flow for two hours, and then emitted water of the colour of blood. Ludovico Aurelio states that the vines, corn, and trees were burnt up over large districts.

19. According to Nicolo Speziale, there was a great eruption from the eastern side of the mountain in 1181.

20. A stream of lava is said to have burst from the eastern side of the mountain in 1285, when Charles of Anjou was on his death-bed, and to have flowed fifteen miles.

21. In the 1329 Niccolo Speziale was in Catania, and witnessed the eruption of which he has left us an account. On the evening of June 28th, about the hour of vespers, Etna was strongly convulsed, terrible noises were emitted, and flames issued from the south side of the mountain. A new crater—Monte Lepre—opened in the Val del Bove above the rock of Musarra, and emitted large quantities of dense black smoke. Soon afterwards a torrent of lava poured from the crater, and red-hot masses of rock were projected into the air. These effects continued till the 15th of July, when a second crater opened ten miles to the S.E. of Montelepre, and near the Church of S. Giovanni Paparometto. Soon after four other craters opened around it, and emitted

smoke and lava. The sun was obscured from morning till evening by the smoke and ashes, and the adjacent fields were burnt up by the hot sand and ashes. Multitudes of birds and animals perished, and many persons are said to have died from terror. The lava streams were divided into three portions, two of which flowed towards Aci, and the third towards Catania. The ashes were carried as far as Malta, a distance of 130 miles.

22. Four years afterwards an eruption is recorded by Silvaggio.

23. A manuscript preserved in the archives of the Cathedral of Catania mentions an eruption which occurred on the 6th of August, 1371, which caused the destruction of numerous olive groves near the city.

24. An eruption which lasted for twelve days commenced on the 9th of November, 1408; it originated in the great crater, but several mouths subsequently opened near the base of the mountain. Large quantities of red-hot ashes were emitted, some of which fell in Calabria. The villages of Pedara and Tre Castagne suffered severely from this eruption.

25. A violent earthquake in 1444 caused the upper cone of the mountain to fall into the crater. A torrent of lava also issued from the mountain, and moved for a space of twenty days towards Catania, but it did not reach the city.

26. Two years later lava issued from the Val del

Bove near the Rock of Musarra; the crater then formed was perhaps the present Monte Finocchio.

27. A short eruption, of which we have no details, occurred in 1447: after which Etna was quiescent for 89 years.

28. Bembo and Fazzello mention an eruption which occurred towards the close of the 15th century, during which a current of lava flowed from the great crater, and destroyed a portion of Catania. In 1533 Filoteo, of whom we have before spoken as one of the earliest historians of Etna, descended into the crater, which possessed its present funnel-like form. He found at the bottom a hole, not larger than a man's head, from which issued a thin moist sulphurous vapour.

29. In March, 1536, a quantity of lava issued from the great crater, and several new apertures opened near the summit of the mountain and emitted lava. It divided into several streams, flowing in different directions, one towards Randazzo, a second towards Aderno, and a third towards Bronte. The lava swept everything before it; at the same time quantities of smoke and ashes were ejected, the mountain was convulsed, and fearful noises were heard. Three new craters were formed on the south and west sides of the mountain, and on the 26th of March twelve new craters, or *bocche*, opened between Monte Manfre and Monte Vituri. A physician of Lentini, named Negro di Piazza, having

approached too near to the scene of the eruption, was destroyed by a volley of red-hot stones. Several rifts were formed in the sides of the mountain from which issued flames and hot cinders.

30. A year later, in May, 1537, a fresh outburst occurred; a number of new mouths were opened on the south slope of the mountain near La Fontanelle, and a quantity of lava was emitted, which flowed in the direction of Catania, destroying a part of Nicolosi, and S. Antonio. In four days the lava had run fifteen miles. At the same time violent shocks of earthquake occurred all over Sicily, the inhabitants thought that the last day had come, and many prepared for their end by receiving Extreme Unction. According to Filoteo the noises were so violent that many persons were struck deaf. The sun was obscured by smoke and dust, ashes fell in sufficient quantities to destroy the olive plantations of Messina, and were even carried 300 miles out to sea. The great crater suddenly fell in, so as to become level with the Piano del Lago. The height of the mountain was thus diminished by 320 feet.

31. Three new craters opened in November, 1566, on the north-east slope of the mountain. Quantities of lava were emitted, which flowed towards Linguaglossa and Randazzo.

32. A slight eruption, of which we have no details, occurred in 1579.

33. According to Carrera, an eruption occurred in June, 1603. The mountain was shaken with earthquakes, and great volumes of smoke and flame were emitted.

34. A stream of lava issued from the great crater four years later, and filled up the lake which had previously existed in the Piano del Lago.

35. In February, 1610, lava was emitted from the great crater. It flowed towards Aderno, and filled up the bed of the Simeto, a little above the Ponte di Carcaci. A few months later a second stream destroyed a large portion of the forest Del Pino.

36. In 1614 several new craters were opened between Randazzo and the great crater on the north side of the mountain. A quantity of lava issued from them, which united into one stream, and ran for ten miles, destroying a great deal of wooded country.

37. A slight eruption occurred in 1619.

38. In February, 1633, Nicolosi was partially destroyed by a violent earthquake; and in the following December earthquakes became frequent on the mountain. A new crater opened above the cone called Serrapizzuta, five miles from the great crater, and emitted a good deal of lava. A second crater afterwards opened about two miles to the east of the former. The eruption lasted off and on for four years: the ejected lava then covered a tract eighteen miles in length by two miles

in width, the thickness sometimes attaining 42 feet. In 1643 a severe earthquake occurred, which was mainly felt on the west side of the mountain.

39. In 1646 a new mouth opened on the north-north-east side of the mountain, five miles from the great crater. The lava flowed towards Castiglione.

40. In February, 1651, several new mouths opened on the west side of the mountain, and poured out vast volumes of lava which threatened to overwhelm Bronte. In twenty-four hours the lava ran sixteen miles with a breadth of four miles.

41. We have a more detailed account of the eruption of 1669 than of any previous outburst. It was observed by many men of different nations; and we find accounts of it in our own *Philosophical Transactions*, in French, and of course in Italian. Perhaps the most accurate and complete description is that given by Alfonso Borelli, Professor of Mathematics in Catania. The eruption was, in every respect, one of the most terrible on record. On the 8th of March the sun was obscured, and a whirlwind blew over the face of the mountain; at the same time earthquakes commenced, and continued to increase in violence for three days, when Nicolosi was converted into a heap of ruins. On the morning of the 11th a fissure nearly twelve miles in length opened in the side of the mountain, and extended from the Piano di S. Leo to Monte Frumento,

a mile from the summit. The fissure was only six
feet wide, but it seemed to be of unknown depth, and
a bright light proceeded from it. Six mouths opened
in a line with the principal fissure ; they emitted vast
columns of smoke, accompanied by loud bellowings
which could be heard 40 miles off. Towards the close
of the day, a crater opened about a mile below the
others, and it ejected red hot stones to a considerable
distance, and afterwards sand and ashes which covered
the country for a distance of 60 miles. The new crater
soon vomited forth a torrent of lava which presented
a front of two miles, it encircled Monpilieri, and afterwards flowed towards Belpasso, a town of 8000 inhabitants, which was speedily destroyed. Seven mouths
of fire opened around the new crater, and in three days
united with it, forming one large crater 800 feet in
diameter. The torrent of lava all this time continued
to descend, and it destroyed the town of Mascalucia on
the 23rd of March. On the same day the crater cast
up great quantities of sand, ashes, and scoriæ, and
formed above itself the great double-coned hill now
called Monti Rossi from the red colour of the ashes of
which it is mainly composed. On the 25th very violent
earthquakes occurred, and the cone of the great central
crater was shaken down into the crater for the fifth
time since the first century A.D. The original current

of lava had divided into three streams, one of which
destroyed S. Pietro, the second Camporotondo, and
the third the lands about Mascalucia, and afterwards
the village of Misterbianco. Fourteen villages were
altogether destroyed, and the lava was on its way to
Catania. At Albanelli, two miles from the city, it
undermined a hill covered with cornfields, and carried
it forward a considerable distance; a vineyard was also
seen to be floating on its fiery surface. When the lava
reached the walls of Catania it accumulated without
progression until it rose to the top of the wall, 60 feet
in height, and it then fell over in a fiery cascade, and
overwhelmed a part of the city. Another portion of the
same stream threw down 120 feet of the wall, and
flowed into the city. On the 23rd of April the lava
reached the sea, which it entered as a stream 600 yards
broad and 40 feet deep. The stream had moved at the
rate of thirteen miles in twenty days, but as it cooled
it moved less quickly, and during the last twenty-three
days of its course it only moved two miles. On reaching
the sea the water of course began to boil violently, and
clouds of steam arose, carrying with them particles of
scoriæ. Towards the end of April the stream on the
west side of Catania, which had appeared to be con-
solidated, again burst forth, and flowed into the garden
of the Benedictine Monastery of S. Niccola, and then

branched off into the city. Attempts were made to build walls to arrest its progress. An attempt of another kind was made by a gentleman of Catania, named Pappalardo, who took fifty men with him, having previously provided them with skins for protection from the intense heat, and with crowbars to effect an opening in the lava. They pierced the solid outer crust of solidified lava, and a rivulet of the molten interior immediately gushed out, and flowed in the direction of Paterno ; whereupon 500 men of that town, alarmed for its safety, took up arms, and caused Pappalardo and his men to desist. The lava did not altogether stop for four months ; and two years after it had ceased to flow it was found to be red hot beneath the surface. Even eight years after the eruption quantities of steam escaped from the lava after a shower of rain. The stones which were ejected from the crater during this eruption were often of considerable magnitude, and Borelli calculated that the diameter of one which he saw was 50 feet ; it was thrown to a distance of a mile, and as it fell it penetrated the earth to a depth of 23 feet. The volume of lava emitted during this eruption amounted to many millions of cubic feet : Ferrara considers that the length of the stream was at least fifteen miles, while its average width was between two and three miles, so that it covered at least forty square miles of surface.

In a somewhat rare tract,[1] Lord Winchelsea, who was returning to England from Constantinople, and who landed at Catania, gives an account of what he saw of the eruption. He appears to have been frightened at the sight, and took good care to keep in a safe place; hence his letter, which is a short one, is mainly founded on hearsay. However, he says, "I could discern the river of fire to descend the mountain, of a terrible fiery or red colour, and stones of a paler red to swim thereon, and to be as big as an ordinary table . . . . Of 20,000 persons which inhabited Catania, 3000 did only remain; all their goods are carried away, the cannon of brass are removed out of the castle, some great bells taken down, the city gates walled up next the fire, and preparations made by all to abandon the city." The noble earl is less happy in his scientific ideas than in his general statement of the facts of which he was an eye-witness; we can only hope that he joined the recently-formed Royal Society on his return to England, and listened to Robert Hooke's discourse on fire. In describing the lava, Lord Winchelsea says, " The

---

[1] " A true and exact relation of the late prodigious earthquake and eruption of Mount Ætna or Monte Gibello; as it came in a letter written to his Majesty from Naples by the Rt. Honble. the Earl of Winchelsea late Ambassadour at Constantinople, who in his return from thence visiting Catania in the Island of Sicily, was an eye-witness of that dreadful spectacle." Published by Authority. Printed by T. Newcomb in the Savoy. 1669.

composition of this fire, stones, and cinders, are sulphur, nitre, quicksilver, sal-ammoniac, lead, iron, brass, and all other mettals!" Two other accounts are appended to the above letter; in one of these we are told that as the lava approached Catania, the various religious bodies carried their relics in procession, "followed by great multitudes of people, some of them mortifying themselves with whips, and other signs of penance, with great complaints and cries, expressing their dreadful expectation of the events of those prodigious fiery inundations." In the midst of all this, news was brought that a large band of robbers had taken advantage of the general distress, and were robbing right and left, and murdering the people: whereupon a troop of Spanish horse was sent out to protect the city and country, three pair of gallows were set up, and such as were found robbing were executed without trial by martial law.

As the lava streams approached the city, the Senate, accompanied by the Bishop and all the clergy, secular and regular, went in procession out of the city to Monte di S. Sofia with all their relics, etc. There they erected an altar in view of the burning mountain, and celebrated mass, "and used the exorcismes accustomed upon such extraordinary occasions, all which time the mountain ceased not as before with excessive roaring to throw up its smoak and flames with extraordinary violence, and abundance of great stones, which were carried through the air."

42. For a few years after this terrible eruption Etna was quiescent, but in 1682 a new mouth opened on the east side of the mountain, a little below the summit, and above the Val del Bove. Lava issued from it, and rushed down the precipices of the Val del Bove as far as the rock of Musarra.

43. Six years later a torrent of lava burst from an opening in the great cone, and flowed into the Val del Bove for a distance of three miles.

44. In the following year lava was emitted from a mouth in the Val del Bove, and it descended for about ten miles, destroying everything in its course, until it reached a little valley near Macchia.

45. Early in January 1693, clouds of black smoke were poured from the great crater, and loud noises resembling the discharge of artillery were heard. A violent earthquake succeeded, and Catania was shaken to the ground, burying 18,000 of its inhabitants in the ruins. It is said that in all fifty towns were destroyed in Sicily, together with from 60,000 to 100,000 inhabitants. Lava was emitted from the crater, which was lowered by the eruption.

46. In the following year Etna again entered into eruption, ejecting large quantities of ashes, some of which were carried as far as Malta.

47. In March 1702, three mouths opened in the Contrada del Trifoglietto, near the head of the Val del Bove. Lava was emitted from them, which flowed into the Valley of Calanna.

48. Towards the end of 1723 loud bellowings issued from the mountain ; earthquakes occurred, and a torrent of lava issued from the crater, which flowed towards Bronte, through the Bosco di Bronte.

49. A small lava stream issued from the crater in 1732, and descended the western slope of the mountain, but without producing any damage.

50. In October 1735, the usual noises which presage an eruption were heard, earthquakes followed, and a little later the crater emitted flames and red-hot stones. Lava also issued from it, and the stream divided into three branches, one of which flowed towards Bronte, a second towards Linguaglossa, and a third towards Mascali ; but they did not get beyond the upper regions of the mountain.

51. In 1744 the mountain threw out great quantities of ashes, but no lava.

52. In 1747 a quantity of lava flowed from the great crater into the Val del Bove, and the height of the cone was considerably increased during the eruption.

53. Early in the year 1755, Etna began to show signs of disturbance ; a great column of black smoke issued from the crater, from which forked lightning

was frequently emitted. Loud detonations were heard, and two streams of lava issued from the crater. A new mouth opened near the Rocca di Musarra in the Val del Bove, four miles from the summit, and a quantity of lava was ejected from it. An extraordinary flood of water descended from the Val del Bove, carrying all before it, and strewing its path, with huge blocks. Recupero estimated the volume of water as 16,000,000 cubic feet, probably a greater amount than could be furnished by the melting of all the winter's snow on the mountain. It formed a channel two miles broad, and, in some places, thirty-four feet deep, and it flowed at the rate of a mile in a minute and a half during the first twelve miles of its course. Lyell considers the flood was probably produced by the melting, not only of the winter's snow, but also of older layers of ice, which were suddenly melted by the permeation of hot steam and lava, and which had been previously preserved from melting by a deposit of sand and ashes, as in the case of the ancient glacier found near the summit of the mountain in 1828. In November 1758, a smart shock of earthquake caused the cone of the great crater to fall in, but no eruption occurred at the time.

54. Great quantities of ashes, and some small streams of lava, were emitted from the crater in 1759, a little later the cone, which had been again raised by

the eruption, gave way, and the greater part of it fell into the crater. Two parts of it however were left standing.

55. Severe shocks of earthquakes were felt on the east side of the mountain in 1763, and a new mouth opened in the Bosco di Bronte, ten miles from the town, between Monte Rosso and Monte Lepre. Four other mouths were afterwards opened in a line; they threw up quantities of scoriæ and ashes, and afterwards lava. In the middle of June several mouths opened on the south side of the mountain, and a fissure 2000 feet long opened downwards in a southerly direction. The lava divided into two branches, the larger of which was ten miles long and 250 feet wide, with a depth of 25 feet.

56. Several new mouths opened in the spring of 1766, and ejected large volumes of ashes, also streams of lava, which flowed in the direction of Nicolosi and Pedara. The Canon Recupero, one of the historians of Etna, witnessed this eruption, and narrowly escaped being destroyed. He had ascended a small hill 50 feet high, of ancient volcanic matter, in order to witness the approach of the lava stream which was slowly advancing with a front of two miles and a half. Suddenly two small streams detached themselves from the main stream, and ran rapidly towards the hill. Recupero and his guide at once hastened to descend,

and had barely escaped when they saw the hill surrounded by lava, and in a few minutes it was melted
down and sank into the molten mass.

57. In the early part of 1780, earthquakes were felt
all over Sicily, and on the 18th of May a fissure opened
on the south-west side of the mountain, and extended
from the base of the great crater for seven miles, ter-
minating in a new mouth from which a stream of lava
emanated. This encountered the cone of Palmintelli
in its course, and separated into two branches, each of
which was 400 feet wide. Other mouths opened later
in the year, and emitted large quantities of lava, which
devastated the country of Montemazzo.

58. In 1781 the volcano emitted a quantity of lava
which flowed into the Val del Bove. Clouds of grey
ashes were also ejected. At the commencement of the
great Calabrian earthquake of 1783, Etna ejected large
quantities of smoke, but it was otherwise quiescent.

59. In the middle of 1787 lava burst from the great
crater, which also discharged quantities of sand, scoriæ,
and red-hot ashes. Large heated masses of rock were
ejected to a great height, and subterranean bellowings
were heard by the dwellers on the mountain.

60. Five years afterwards a fresh outburst occurred,
earthquakes were prevalent, and vast volumes of smoke
bore to seaward, and seemed to bridge the sea between
Sicily and Africa. A torrent of lava flowed towards

Aderno, and a second flowed into the Val del Bove as far as Zoccolaro. A pit called *La Cisterna*, 40 feet in diameter, opened in the Piano del Lago, near the great cone, and ejected smoke and masses of old lava saturated with water. Several mouths opened below the crater, and the country round about Zaffarana was desolated. The Abate Ferrara, the author of the *Descrizione dell' Etna*, witnessed this eruption: "I shall never forget," he writes, "that this last mouth opened precisely on the spot where, the day before, I had made my meal with a shepherd. On my return next day he related how, after a stunning explosion, the rocks on which we had sat together were blown into the air, and a mouth opened, discharging a flood of fire, which, rushing down with the rapidity of water, hardly gave him time to make his escape."

61. In 1797 a slight eruption occurred, and the great crater threw out ashes and sand, but no lava. Earthquakes were frequent.

62. In the following year lava was emitted, and severe earthquakes occurred.

63. The eruptions continued during 1799.

64. In February 1800 loud explosions were heard by the dwellers on the mountain, and columns of fire issued from the crater, accompanied by forked lightning. This was succeeded by a discharge of hot ashes and scoriæ, which, falling on the snows accumulated near

the summit of the mountain, produced devastating floods of water.

65. In November 1802 a new mouth opened near the Rocca di Musarra in the Val del Bove, which emitted a copious stream of lava. In a day and a half the lava had run twelve miles.

66. In 1805 the great crater was in a state of eruption, and a cone was thrown up within it to a height of 1,050 feet.

67. In 1808 the mountain again became active, and fire and smoke were emitted from the crater.

68. In March 1809, no less than twenty-one mouths of fire opened in the direction of Castiglione. They ejected volumes of smoke, large quantities of scoriæ and ashes, and afterwards lava, which, uniting into one torrent, flowed with a front of 450 feet for 8 miles. Fissures were formed in the earth, and loud explosions constantly occurred within the great crater; a small cone was thrown up.

69. Two years afterwards more than thirty mouths opened in a line running eastwards for five miles. They ejected jets of fire accompanied by much smoke. The eruptions soon diminished in the higher mouths, and became more and more violent in the lower mouths, until the eruption centred in the lowest one called S. Simone, near the head of the Val del Bove. From this, great black clouds, having a lustre like that of black

wool, issued, and afterwards quantities of lava, which formed a stream a mile wide, and eight miles long. It flowed nearly as far as the village of Milo. Frequent earthquakes accompanied this outburst, and they continued in various parts of the island for the following five years.

70. In 1819 five new mouths of fire opened near the scene of the eruption of 1811; three of these united into one large crater, and poured forth a quantity of lava into the Val del Bove. The lava flowed until it reached a nearly perpendicular precipice at the bend of the valley of Calanna, over which it fell in a cascade, and, being hardened by its descent, it was forced against the sides of the tufaceous rock at the bottom, so as to produce an extraordinary amount of abrasion, accompanied by clouds of dust, worn off by the friction. Mr. Scrope observed that the lava flowed at the rate of about a yard an hour, nine months after its emission.

71. A slight eruption occurred in 1831 from the great crater, which threw out lava on its northern side.

72. In October of the following year a violent eruption occurred. A new crater was formed in the Val del Serbo, above Bronte and three miles from the summit. Seven mouths afterwards opened, three miles below the first. From one of these lava was emitted, which flowed to within a mile and a half of Bronte. The stream was a mile and a half broad, and 40 feet deep.

73. A slight eruption occurred in 1838, when a small quantity of lava was poured from the great crater into the Val del Bove.

74. Four years later the crater discharged ashes and scoriæ, and lava burst from the cone 300 feet from the summit. It flowed into the Val del Bove, in a stream 600 feet wide, and it came to a standstill ten miles from the summit.

75. Near the end of the following year, fifteen mouths of fire opened near the crater of 1832, at a height of 7000 feet above the sea. They began by discharging scoriæ and sand, and afterwards lava, which divided into three streams, the two outer ones soon came to a standstill, while the central stream continued to flow at the rapid rate of 180 feet a minute, the descent being an angle of 25°. The heat at a distance of 120 feet from the current was 90° F. A new crater opened just above Bronte, and discharged lava which threatened the town, but it fortunately encountered Monte Vittoria and was diverted into another course. While a number of the inhabitants of Bronte were watching the progress of the lava, the front of the stream was suddenly blown out as by an explosion of gunpowder; in an instant red-hot masses were hurled in every direction; and a cloud of vapour enveloped everything. Thirty-six persons were killed on the spot, and twenty survived but a few hours. The great crater showed signs of disturbance, by emit-

ting dense volumes of smoke, and loud bellowings, also quantities of volcanic dust saturated with hydrochloric acid, which destroyed the vegetation wherever it fell.

76. A very violent eruption which lasted more than nine months, commenced on the 21st of August, 1852. It was first witnessed by a party of six English tourists, who were ascending the mountain from Nicolosi in order to see the sunrise from the summit. As they approached the Casa Inglesi the crater commenced to give forth ashes and flames of fire. In a narrow defile they were met by a violent hurricane, which overthrew both the mules and their riders, and urged them towards the precipices of the Val del Bove. They sheltered themselves beneath some masses of lava, when suddenly an earthquake shook the mountain, and their mules in terror fled away. They returned on foot towards daylight to Nicolosi, fortunately without having sustained injury. In the course of the night many *bocche del fuoco* opened in that part of the Val del Bove called the Balzo di Trifoglietto, and a great fissure opened at the base of the Giannicola Grande, and a crater was thrown up from which for seventeen days showers of sand and scoriæ were ejected. During the next day a quantity of lava flowed down the Val del Bove, branching off so that one stream advanced to the foot of Monte Finocchio, and the other to Monte Calanna. Afterwards it flowed towards Zaffarana, and devastated a large tract

of woody region. Four days later a second crater was formed near the first, from which lava was emitted together with sand and scoriæ, which caused cones to rise around the craters. The lava moved but slowly, and towards the end of August it came to a stand, only a quarter of a mile from Zaffarana : on the second of September, Gemellaro ascended Monte Finocchio in the Val del Bove in order to witness the outburst. He states that the hill was violently agitated, like a ship at sea. The surface of the Val del Bove appeared like a molten lake; scoriæ were thrown up from the craters to a great height, and loud explosions were heard at frequent intervals. The eruption continued to increase in violence. On October 6th two new mouths opened in the Val del Bove, emitting lava which flowed towards the Valley of Calanna, and fell over the Salto della Giumenta, a precipice nearly 200 feet deep. The noise which it produced was like that of the clash of metallic masses. The eruption continued with abated violence during the early months of 1853, and it did not finally cease till May 27. The entire mass of lava ejected is estimated to be equal to an area six miles long by two miles broad, with an average depth of about twelve feet.

I am indebted to M. Antonin Moris of Palermo for the following account of the eruption of 1852 :

The eruption of 1852 commenced on the 21st of August. The earthquakes, the jets of flame from the great crater, and the subterranean rumblings which usually precede an eruption, did not herald the approach of this one. An English family, who were then making the ascent of the mountain, together with a poor shepherd of Riposto, were the only witnesses of the first outburst. The latter was asleep in the midst of his flocks, and was awakened by violent shakings of the ground; he fled in haste, and some seconds afterwards the earth opened with a loud noise, vomiting a terrible column of fire, at the very spot which he had just abandoned. An enormous crevasse opened on the north side of Trifoglietto in the direction of the great crater. On its summit near the opening called the Piccolo Teatro, several openings were produced at the very first, but they only emitted feeble currents of lava. All the force of the eruption was concentrated at the foot of the escarpment of the Serra di Giannicola, 4 kilometres, (2½ miles) from the summit of Etna. To the west of, and somewhat above the principal crater, a second one was formed, but its activity was of short duration. The liquid lava issued with such violence that in 24 hours it had reached the base of Monte Calanna, a distance of 3 kilometres, (nearly 2 miles). After surrounding this hill, it divided into two currents, one of which ran towards Zaffarana, and the other towards Milo. At a distance they seemed to present a united front of 2 kilometres, (1¼ mile), which threatened to destroy all

the villages below. The Val del Bove was already
entirely overrun; Isoletta dei Zappinelli in the midst
of the lavas of 1811 and 1819 was overwhelmed; the
valley of Calanna was buried under the fire with lava,
when on the 28th of August the lava hurled itself into
the narrow passage of the Portella di Calanna. A
frightful cascade of lava was then seen to precipitate
itself from a height of 60 metres, with a harsh metallic
noise, accompanied by loud cracking. Zaffarana was on
the eve of total ruin; the fire had taken the direction of
the ravine which terminates there, when suddenly, in
the beginning of September, the devastating stream
stayed its march against the ill-fated district.

On the contrary that which had taken the direction of
Milo, reinforced by a new current on the 10th of September, destroyed the hamlet of Caselle del Milo; and
afterwards divided itself into two branches, which left
the village of Caselle in safety between them.

The inhabitants of La Macchia and Giarre gave themselves up for lost; for it seemed that the lava would be
obliged to follow the valley of Santa Maria della Strada;
happily, however, from the 20th of September onward,
it ceased to advance perceptibly. The eruption did not
totally subside till March 1853; but the lava-flows did
no more than travel by the side, or on the top of the
older, without extending beyond them.

The crater of 1852 was called the Centenario, from its

having been formed at the time of the centenary of the fête of S. Agatha. Santiago, in the island of Cuba, was destroyed by an earthquake on the very day of the eruption.

During the whole period of the eruption, only one explosion proceeded from the great crater of Etna. By it an enormous column of ashes and scoriæ was cast into the air.

On the 9th of September white ashes were seen on the summit, which at a distance appeared like snow. When pressed together by the hand they took the consistence of clay, but they hardened in the fire, and could then be reduced to powder. They have been considered to be the *debris* of felspathic rocks, disintegrated by the heat of the lava, and blown out by the expansive power of disengaged gas.

The eruption of 1852 was one of the grandest of the recorded eruptions of Etna. More than 2,000,000,000 cubic feet of red hot lava was spread over three square miles. This eruption was minutely described by Carlo Gemellaro, in a memoir entitled, "*Breve ragguaglio della eruzione dell' Etna del 21 Agosto, 1852.*"

77. In October 1864, frequent shocks of earthquakes were felt by the dwellers on Etna. In January clouds of smoke were emitted by the great crater, and roaring sounds were heard. On the night of the 30th a violent shock was felt on the north-east side of the mountain,

and a mouth opened below Monte Frumento, from which lava was ejected. It flowed at a rate of about a mile a day, and ultimately divided into two streams. By March 10th the new mouths of fire had increased to seven in number, and they were all situated along a line stretching down from the summit. The three upper craters gave forth loud detonations three or four times a minute. Professor Orazio Silvestri has devoted a quarto of 267 pages to an account of *I Fenomeni Vulcanici presentati dall' Etna nel 1863-64-65-66.*

78. In August 1874 the inhabitants of the towns situated on the north, west, and east sides of the mountain, were awakened by loud subterranean rumblings. Soon afterwards a formidable column of black smoke issued from the crater, accompanied by sand, scoriæ, and ignited matter (*infuocata materia*). Severe shocks of earthquake were felt, the centre of impulsion being apparently situated on the northern flank of the mountain, at a height of 2450 metres above the level of the sea. Some small *bocche eruttive* opened near the great crater, and ejected lava, but the quantity was comparatively small, and but little damage was done. An account of this eruption was given by Silvestri in 1874, in a small pamphlet entitled, *Notizie sulla eruzione dell' Etna del 29 Agosto, 1874.* Since 1874 the mountain has been in a quiescent state. The centre of disturbance was at an elevation of 2450 metres (7600 feet)

above the sea, on the north side of the crater, and between the minor cones known as the *Fratelli Pii* and *Monte Grigio*. A new crater, having an elliptical contour, and a diameter of about 100 metres, was formed at this point. It is composed of a prehistoric grey labradorite, and of doleritic lava. Downwards from the main crater, in the direction of Monte di Mojo, a long fissure extended for 400 metres, and along the line of this fissure no less than *thirty-five* minor cones opened, with craters of from thirty to three metres in diameter. The stream of lava ejected from the various *boccarelle* was 400 metres long, 80 wide, and 2 metres in thickness, and the bulk of volcanic material brought to the surface, including the principal cone and its thirty-five subordinates and their ejectamenta, was calculated to amount to 1,351,000 cubic metres. The lava is of an augitic character, and magnetic; it possesses a specific gravity of 2·3636 at 25° C.

It will be seen from the account of the foregoing eruptions that there is a great similarity in the character of the eruptions of Etna. Earthquakes presage the outburst; loud explosions follow, rifts and *bocche del fuoco* open in the sides of the mountain; smoke, sand, ashes, and scoriæ are discharged, the action localises itself in one or more craters, cinders are thrown up, and accumulate around the crater and cone, ultimately lava rises, and frequently breaks down one side of the cone,

where the resistance is least. Then the eruption is at
an end.

Smyth says, "The symptoms which precede an eruption are generally irregular clouds of smoke, ferilli, or
volcanic lightnings, hollow intonations, and local earthquakes that often alarm the surrounding country as far
as Messina, and have given the whole province the name
of Val Demone, as being the abode of infernal spirits.
These agitations increase until the vast cauldron becomes
surcharged with the fused minerals, when, if the convulsion is not sufficiently powerful to force them from
the great crater (which, from its great altitude and the
weight of the candent matter, requires an uncommon
effort), they explode through that part of the side which
offers the least resistance with a grand and terrific
effect, throwing red-hot stones and flakes of fire to an
incredible height, and spreading ignited cinders and
ashes in every direction." After the eruption of ashes,
lava frequently follows, sometimes rising to the top of
the cone of cinders, at others breaching it on the least
resisting side. When the lava has reached the base
of the cone, it begins to flow down the mountain, and
being then in a very fluid state, it moves with great
velocity. As it cools the sides and surface begin to
harden, its velocity decreases, and in the course of a
few days it only moves a few yards in an hour. The
internal portions, however, part slowly with their heat,

and months after the eruption, clouds of steam arise from the black and externally cold lava beds after rain, which, having penetrated through the cracks, has found its way to the heated mass within.

Of the seventy-eight eruptions described above, it will be noticed that not more than nineteen have been of extreme violence, while the majority have been of a slight and comparatively harmless character.

CHAPTER VI.

GEOLOGY AND MINERALOGY OF THE MOUNTAIN.

Elie de Beaumont's classification of rocks of Etna.—Hoffman's geological map.—Lyell's researches.—The period of earliest eruption.—The Val del Bove.—Two craters of eruption.—Antiquity of Etna.—The lavas of Etna.—Labradorite.—Augite.—Olivine.—Analcime.—Titaniferous iron.—Mr. Rutley's examination of Etna lavas under the microscope.

The opinion of geologists is divided as to the manner in which a volcano is first formed. Some hold that the volcanic forces have upraised the rocks from beneath, and at last finding vent have scattered the lighter portions of such rocks into the air, and have poured out lava through the rent masses, thus forming a *crater of elevation*. Others maintain that the volcanic products are ejected from an aperture or fissure already existing in rocks previously formed, and that the accumulation of these products around the vent forms the mass of the volcano and the *crater of eruption*. Lyell favours the latter view; Von Buch, Dufrénoy, and Elie de Beaumont the former.

ETNA
AND ITS ENVIRONS

English Miles

N O I

Catunia

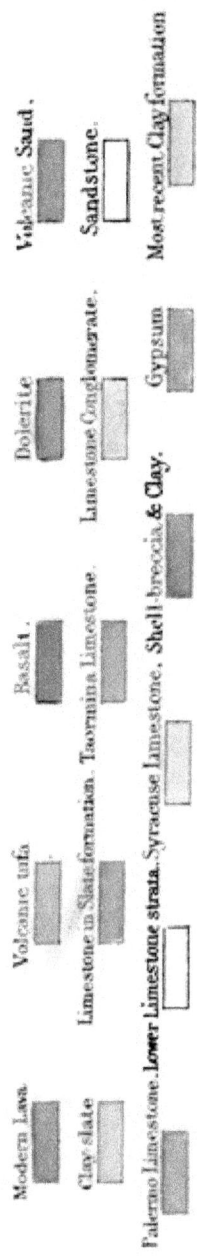

Modern Lava. Volcanic tufa. Basalt. Dolerite. Volcanic Sand.

Clay slate. Sandstone.

Palermo Limestone. Lower Limestone strata. Limestone in Slate formation. Taormina limestone. Limestone Conglomerate.

Syracuse limestone. Shell-breccia & Clay. Gypsum. Most recent Clay formation

According to M. Elie de Beaumont, Etna is an irregular crater of elevation. The original deposits were nearly horizontal, and lavas were poured through fissures in these, and accumulated at first in layers; afterwards the whole mass was upheaved and a crater formed.[1] The upheaving force does not appear to have acted at one point, but along a line traversing the Val del Bove. The latter he refers to a subsidence of a portion of the mountain. He divides the rocks of Etna into six orders: 1. The lowest basis of the mountain would appear to consist of granite, because masses of that rock have from time to time been ejected. 2. Calcareous and arenaceous rocks, of which the mountains surrounding Etna are composed, and which appear capped with lava near Bronte and elsewhere. 3. Basaltic rocks, which are met with near Motta S. Anastasia, Paterno, Licodia, and Aderno, and in the Isole de' Ciclopi. 4. Rolled pebbles, which form a range of slightly rising ground between the first slopes of Etna on the southern side and the plain of Catania. (Lyell speaks of this rising ground as consisting of "argillaceous and sandy beds with marine shells, nearly all of living Mediterranean species, and with associated and contemporaneous volcanic rocks.") 5. Ancient lavas forming the escarpments around the Val del Bove; and 6th,

[1] " Récherches sur la structure et sur l'origine du Mont Etna." 1836.

Modern lavas. He considers that the fissures which
abound on Etna are shifts or faults produced by dis-
location, and that the minor cones are points along such
fissures from which ashes and lava have been ejected.
He admits the existence of two cones. The geological
map of Etna prepared by M. Elie de Beaumont to
accompany his memoir can scarcely be regarded as a
great addition to our knowledge of the mountain. For
although in the main points it is correct, so many details
have been omitted that the map must be considered
to have now been quite superseded by those of Von
Waltershausen and Friedrich Hoffmann.

The most convenient geological map of the mountain
is without doubt that of Hoffmann, given in the
Vulkanen Atlas of Dr. Von Leonhard; and here repro-
duced. Von Waltershausen's geological map has been
the foundation of all others which have subsequently
appeared. It is a marvel of accurate work, and patient
industry. The form however is inconvenient, as it no-
where appears as a whole, but in separate portions, which
are scattered through the folio sheets of the very expen-
sive *Atlas des Aetna.* It is accurate, and at the same
time very clear and intelligible. By reference to the
map it will be seen that from Capo di Schiso westward,
to near Paterno, Etna is surrounded by sandstone hills ;
at the south we have recent clays, and, at intervals,
chalk. A large triangular space having the two angles

Map of the Val del Bove, to illustrate the theory of a double axis of eruption. (*Lyell*).

at its base, respectively near Maletto and Aderno, and its apex at the great crater, is covered with new lava; while around Nicolosi there is volcanic sand. At the Isole de' Ciclopi, Motta S. Anastasia, and a few other places, basalt is seen; on each side of the Val del Bove, dolerite; and near Misterbianco and Piedemonte, small deposits of clay slate. The great mass of the surface of the mountain, not specially mentioned above, is volcanic tuff.

Among the more important and recent additions to our knowledge of the geology of Etna may be mentioned Lyell's paper on the subject, communicated to the Royal Society in 1858, the matter of which is incorporated in a lengthy chapter on Etna in the "Principles of Geology." Lyell visited the mountain in 1828, 1857, and 1858, and he then collected together a great number of personal observations; he also made use of the maps and plans of Von Waltershausen, and he has analysed the views of Elie de Beaumont and other writers. He alludes at the outset to the numerous minor cones of Etna produced by lateral eruption, and points out the fact that they are gradually obliterated by the lava descending from the upper part of the mountain, which flows around them and heightens the ground on which they stand. In this way the crater of Monte Nocilla is now level with the plain, and the crater of Monte Capreolo was nearly filled by a lava stream in

Ideal section of Mount Etna, from West 35° N. to East 35° S., to illustrate the theory of a double axis of eruption. See M. N. Map, p. 117 (*Lyell*).

A. Axis of Mongibello.

B. Axis of Trifoglietto.

a', c, b', i, d. Older lavas, chiefly trachytic.

c, e, and *d, f.* Lavas chiefly doleritic, poured out from A after the axis or focus B was spent, and before the origin of the Val del Bove.

gg. Scoriæ and lavas of later date than the Val del Bove.

h, i, k. Val del Bove. The faint lines represent the missing rocks.

N.B. In the section between *i* and *k*, it will be seen that the beds at the base, or near *i*, dip steeply away from the Val del Bove: those in the middle, or below *k*, are horizontal, and those at top, or at *k*, dip gently towards the Val del Bove.

L. Older tertiary and sandy rocks, chiefly sandstones.

1669. Thus without doubt beneath the sloping sides of Etna a multitude of obliterated monticules exist.

The strata which surround Mount Etna on the south are of Newer Pliocene date, and contain shells which are nearly all of species still living in the Mediterranean. Out of sixty-five species collected by Lyell in 1828, sixty-one were found to belong to species still common in the Mediterranean. These strata are about the age of the Norwich crag; and the oldest eruptions of Etna must have taken place during the glacial period, but before the period of greatest cold in Northern Europe.

Before visiting Etna, Lyell had been told by Dr. Buckland that in his opinion the Val del Bove was the most interesting part of Etna, accordingly he specially and minutely examined that part of the mountain. This vast valley is situated on the eastern flanks of the mountain, and it commences near the base of the cone, stretching for nearly five miles downwards. It is a large oval basin formed in the side of the mountain, and surrounded by vast precipices, some of which at the head of the valley are between three and four thousand feet in height. The surface is covered with lava of various dates, and several minor cones, notably those of 1852, are within its boundaries. The abrupt precipices reveal the presence of a large number of vertical dikes, radiating from a point within the valley, some of them, according to Von Walter-

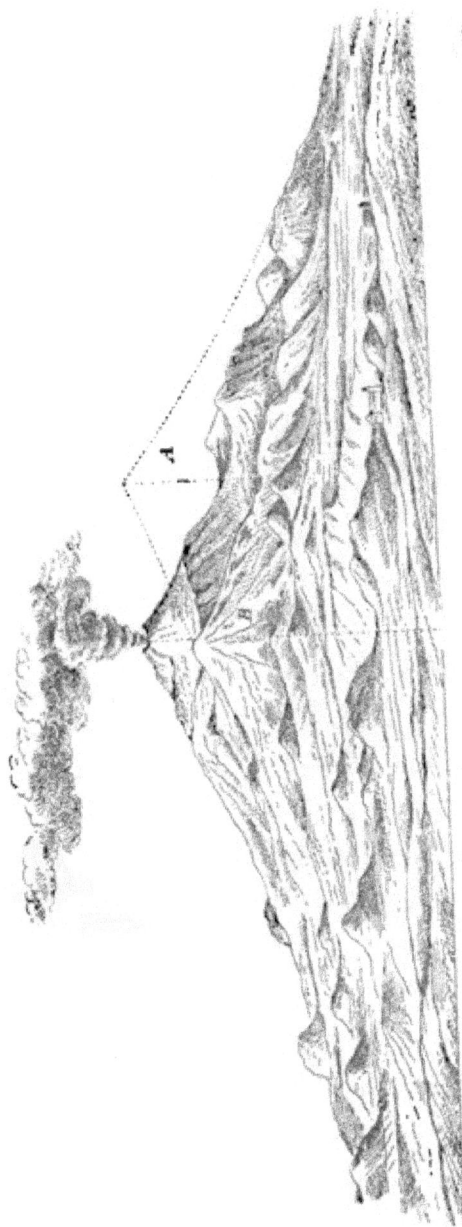

Profile of Etna. A. The ancient mountain, chiefly composed of felspathic rocks. B. The modern mountain, chiefly composed of pyroxenic rocks. (From Gemellaro's *Vulcanologia dell' Etna*).

shausen, being of ancient greenstone. Other dikes of more modern doleritic lava radiate from the present crater. From the slope of the beds in the **Val del Bove**, Lyell and Von Waltershausen have independently inferred that there was once a second great centre of eruption in the Val del Bove between the Sierra Giannicola, and Zoccolaro (*vide* the Figure on p. 117). The axis of eruption passing through this point Lyell calls the *Axis of Trifoglietto;* while he distinguishes the present centre of eruption as the *Axis of Mongibello.* These centres probably existed simultaneously, but were unequal as regards eruptive violence; the crater of Mongibello was the more active of the two, and eventually overwhelmed the crater of Trifoglietto with its products, by which means the whole mountain became a fairly symmetrical cone, having the crater of Mongibello at its apex (*vide* the Figures on pp. 119 and 121). Subsequently the Val del Bove was formed, probably by some paroxysmal explosion, caused by pent-up gases escaping from fissures. Possibly also subsidence may have occurred.

We must then in the first place think of Etna as a submarine volcano of the Newer Pliocene age; when it reached the surface it increased rapidly in bulk by pouring out scoriæ and lava from its two centres of eruption—the centre of Mongibello, and the centre of Trifoglietto,—general upheaval of the surrounding district followed, and ultimately the crater of Trifoglietto was obliterated by the discharges from the crater of Mongibello. Afterwards the Val del Bove was blown out by sudden eruptive force from beneath, and the mountain assumed its present aspect. Then the historical eruptions commenced, and of these we have given an account in the preceding chapter.

The most obvious method of obtaining some idea as to the age of Etna, is to ascertain the thickness of matter added during the historical period to the sides of the mountain, and to compare this with the thickness of the beds of ancient lava and scoriæ exposed at the abrupt precipices of the Val del Bove. There is reason for believing, however, that none of the ancient lavas equalled in volume the lava streams of 1809 and 1852, and the question is much complicated by other considerations. Lyell compares the growth of a volcano to that of an exogenous tree, which increases both in bulk and height by the external application of ligneous matter. Branches which shoot out from the trunk, first pierce the bark and proceed outwards, but if they die or are broken off they become inclosed in the body of the tree, forming knots in the wood. Similarly the volcano consists of a series of conical masses placed one above the other, while the minor cones, corresponding to the branches of the tree, first project, and then become buried again, as successive layers of lava flow around them. But volcanic action is very intermittent, the layers of lava and scoriæ do not accumulate evenly and regularly like the layers of a tree. A violent paroxysmal outbreak may be succeeded by centuries of quiescence, or by a number of ordinary eruptions; or, again, several paroxysmal outbreaks may occur in succession. Moreover, each conical envelope of the mountain is made up of a number of distinct currents of lava, and showers of scoriæ. "Yet we cannot fail to form the most exalted conception of the antiquity of this mountain, when we consider that its base is about 90 miles in circumference; so that it would require ninety flows of lava, each a mile in breadth at their termination, to raise the present foot of the volcano as much as the average height of one lava current." If all the minor cones now visible on Etna could be removed, with all the lava and scoriæ which have ever proceeded from them, the mountain would appear scarcely perceptibly smaller. Other cones would reveal themselves beneath those now existing. Since the time when, in the Newer Pliocene period, the foundations of Etna were laid in the sea, it is quite impossible even to hint at the number of hundreds of thousands of years which have elapsed.

We collected specimens of lava from various points around and upon the mountain. They presented a wonderful similarity of structure, and a mineralogist to whom they were shown remarked that they might almost all have come from the same crater, at the same time. A specimen of the lava of 1535 found near Borello, was ground by a lapidary until it was sufficiently transparent to be examined under the microscope by polarised light. It was found to contain good crystals of augite and olivine, well striated labradorite, and titaniferous iron ore.

Elie de Beaumont affirms that the lavas of Etna consist of labradorite, pyroxene (augite), peridot (olivine), and titaniferous iron. Rose was the first to prove that the lavas of Etna do not contain ordinary felspar (or potash felspar), but labradorite (or lime felspar.) (*Annales des Mines*, 3 serie, t. viii., p. 3.) Elie de Beaumont detached a quantity of white crystals from the interior of a lava found between Giarre and Aci Reale ; these were analysed by M. Auguste Laurent with the following results in 100 parts :—

Silica	47·9
Alumina	34·0
Oxide of Iron	2·4
Soda (Na_2O)	5·1
Potash (K_2O)	·9
Lime	9·5
Magnesia	·2
	100·0

Von Waltershausen gives the following as the composition of two specimens of Labradorite from Etna :—

	I.	II.
Silica	53·56	55·83
Alumina	25·82	25·31
Sesquioxide of Iron	3·41	3·64
Magnesia	·52	·74
Lime	11·69	10·49
Soda	4·09	3·52
Potash	·54	·83
Water	·95	—
	100·58	100·36

Specimens of Augite from Etna have been examined by Von Waltershausen and Rammelsberg, with the following results:—

	Black.	Greenish Black.	From Mascali.	From Monti Rossi.
Silica	47·63	51·70	49·69	47·38
Alumina	6·74	4·38	5·22	5·52
Protoxide of Iron	11·39	4.24	10·75	7·89
,, Manganese	·21	—	—	·10
Magnesia	12·90	21·11	14·74	15·29
Lime	20·87	18·02	18·44	19·10
Sesquioxide of Iron	—	—	—	3·85
Water	·28	·49	·51	·43
	100·02	99·94	99·35	99·56

Olivine is generally met with in the lavas of Etna. It has an olive, or bottle-glass green colour, and is disseminated through the lavas in the form of small crystalline grains, sometimes of some magnitude.

Specific gravity 3·334. A specimen from Etna gave the following results on analysis:—

Silica	41·01
Protoxide of Iron	10·06
Magnesia	47·27
Alumina	·64
Oxide of Nickel	·20
Water	1·04
	100·22

The titaniferous iron of Etna is found disseminated through the mass of the lavas, and is plainly distinguished when a thin section is examined under the microscope. It is sometimes met with in masses. A specimen from Etna, analysed by Von Waltershausen, was found to contain :—

Titanic Acid	11·14
Sesquioxide of Iron	58·86
Protoxide of Iron	30·00
	100·00

The basalts of the Isole de' Ciclopi enclose beautiful transparent crystals of Analcime, the *zeolite dure* of Dolomieu. The word is derived from αναλχις weak, in allusion to the weak electric power which the mineral acquires when heated or rubbed. Dana prefers the time *analcite*. Specimens from the Cyclops Islands have been analysed by Von Waltershausen and Rammelsberg, with the following results:—

	I.	II.	III.
Silica	53·72	55·22	54·34
Alumina	24·03	23·14	23·61
Lime	1·23	·25	·21
Soda (Na₂ O)	7·92	12·19	12·95
Potash (K₂ O)	4·46	1·52	·66
Water	8·50	7·68	8·11
Magnesia	·05	—	—
Sesquioxide of Iron	—	—	·12
	99·91	100·00	100·00

The minerals of Etna are not nearly as numerous as those of Vesuvius. It has been remarked that no area of equal size on the face of the globe furnishes so many different species of minerals as Vesuvius and its immediate neighbourhood. Out of the 380 species of simple minerals enumerated by Hauy, no less than 82 had been found on and around Vesuvius, as long ago as 1828, and many have been since found.

Of other common products of Etna, there are sulphur in various forms, sulphurous acid gas, ammonia salts, hydrochloric acid gas, and steam. A curious white mass, which we found near the summit, proved to be the result of the decomposition of lava by hot acid vapours. In the different lavas, the crystals of labradorite, and of olivine, vary in size considerably. Magnetic oxide of iron is very visible in thin slices of the lavas when placed under the microscope ; and iron appears to be a constant constituent in nearly all the products of the mountain. Within the last few months Prof. Silvestri has detected a mineral oil in the cavities of a prehistoric doleritic lava found near Paterno.[1] The lava is in close contiguity to the clay deposits of a mud volcano, and when examined under the microscope is seen to consist mainly of augite, together with olivine and transparent crystals of labradorite. It contains numerous cavities coated with arragonite, and filled with a mineral oil which constitutes about one per cent of the whole weight of the lava. It was taken from the lava at a temperature of 24° C., (75·2° F.), and solidified at 17° C. (62·6° F.) to a yellowish green mass, which on analysis gave the following percentage composition :—

Liquid hydrocarbons boiling at 79° C.	= 17·97
Hydrocarbons solidifying below 0° C., boiling between 280° and 400° C.	= 31·95
Paraffine melting between 52° and 57° C.	= 42·79
Asphalt containing 12 per cent of ash	= 2·90
Sulphur	= 4·32
	99·93

Prof. Silvestri has recently made some interesting determinations of the specific gravity and chemical

---

[1] "Atti Accademia Gioenia," serie iii., vol. xii.

composition of the different products of Etna. They are given in full in his work entitled, "*I Fenomeni Vulcanici presentati dall' Etna, nel* 1863, 1864, 1865, 1866," which was published in Catania in 1867. The following table gives the specific gravity of various ancient and modern forms of lava, ashes, etc. of Etna :—

	Sp. Gr.
Ashes ejected in 1865	2·644
Sand ,, ,, ,,	2·715
Scoriæ ,, ,, ,,	2·633
Compact lava ,, ,,	2·771
Scoriæ ejected in 1669	2·622
Compact lava ,, ,,	2·697
Lapilli ejected in 1444	2·420
Compact lava ejected in prehistoric times	2·854

A very decided change in the specific gravity was found to take place after fusion. This can only be accounted for on the supposition that a chemical change is effected during the fusion :—

	Sp. Gr. before fusion.	Sp. Gr. after fusion.
Pyroxene of Etna	3·453	2·148
Felspar ,, ,,	2·925	1·361
Olivine ,, ,,	3·410	2·290
Lava of 1865	2·771	1·972
Ancient basaltic lava from the Scogli de' Ciclopi	2·854	2·000
Ancient basaltic lava from Aci Reale	2·795	1·947

It will be seen from the following analyses that the sand, ashes, scoriæ, and compact lava have virtually the same composition—indeed they consist of the same substance in different states of aggregation.

	Ashes.	Sand.	Scoriæ.	Compact lava.
Silica	50·00	49·80	50·00	49·95
Alumina	19·08	18·20	19·00	18·75
Protoxide of iron	12·16	12·42	11·70	11·21
Protoxide of Manganese	·40	·45	·50	·49
Lime	9·98	11·00	10·28	11·10
Magnesia	4·12	4·00	4·20	4·05
Potash	·60	·49	·69	·70
Soda	3·72	3·60	3·40	3·71
Water	·36	·29	·33	·23
Phosphoric acid Titanic acid Vanadic acid Sesquioxide of iron	traces	traces	traces	traces
	100·42	100·25	100·10	100·19

With these we may compare the composition of the lava which issued from Monti Rossi in 1669, and was analysed by Lowe, and of an ancient lava of Etna ejected during an unknown eruption, and analysed by Hesser.

	Ancient lava.	*Lava of* 1669.
Silica	49·63	48·83
Alumina	22·47	16·15
Protoxide of Iron	10·80	16·32
Protoxide of Manganese	·63	·54
Lime	9·05	9·31
Magnesia	2·68	4·58
Soda	3·07	3·45
Potash	·98	·77
	99·31	99·95

The sublimations from the fumaroles are chiefly chloride of ammonium, perchloride of iron, and sulphur. An analysis of the gases of the fumaroles of 1865 gave the following results :—

Carbonic acid	50·5
Hydrosulphuric acid	11·9
Oxygen	7·1
Nitrogen	30·5
	100·0

An account of microscopic analysis of some of the lavas of Etna, for which I am indebted to Mr. Frank Rutley, will be found appended to this chapter. He considers that they are Plagioclase-basalts, and occasionally Olivine-basalts ; and that they consist of Plagioclase, Augite, Olivine, Magnetite, Titaniferous iron, and a residuum of glass.

Near the summit of the great crater I found a mass of perfectly white, vesicular, and very friable substance, somewhat pumiceous in appearance. It proved to be a decomposed lava, and was found elsewhere on the sides of the crater. Mr. Rutley examined a section of it, and reports: "Under the microscope a tolerably thin section shows the outlines of felspar crystals, lying in a hazy milk-white semi-opaque granular matrix. The felspar crystals are lighter and more translucent than the matrix, but are of much the same character, having a granulated or flocculent appearance, somewhat like that of the decomposed felspars in diabase. There are numerous roundish cavities in the section which may once have contained olivine, or some other mineral, or they may be merely vesicles."

A qualitative analysis of this substance, made by Mr. H. M. Elder, has proved that it contains a large quantity of Silica (about 70 per cent.), and smaller proportions of Alumina, Iron, Magnesium, Calcium, and Potash; together with very small amounts of Sulphuric Acid and a trace of Ammonia. Lithium is absent, and Sodium is only present in very minute quantity. Water is present to the extent of nearly 20 per cent.

During the eruption of Etna in 1869 Von Waltershausen noticed on some of the lava blocks which were still hot and smoking, silver-coloured particles, which rapidly underwent change. An insufficient quantity for analysis was collected, but during the eruption of 1874, Silvestri found a quantity of the substance and analysed it. (*Poggendorff's Annalen*, CLVII. 165, 1876.) It possesses a specific gravity of 3·147, and shows a metallic lustre similar to that of steel. On analysis it was found to consist of :—

Iron	90·86
Nitrogen	9·14
	100·00

which corresponds with the formula Fe_5N_2,—a formula assigned by Fremy to Nitride of iron. It has been named *Siderazote*. This new mineral species appears to be formed by the action of hydrochloric acid, and of ammonia on red-hot lava containing a large percentage of iron. It was formed artificially by exposing fragments of lava alternately to the action of hydrochloric acid and ammonia in a red-hot tube. At a high temperature Siderazote undergoes decomposition, nitrogen being evolved. In contact with steam at a red heat it forms magnetite and ammonia.

The Mineral Constitution and Microscopic Characters of some of the Lavas of Etna.

By Frank Rutley, f.r.g.s., of H.M. Geological Survey.

A cursory examination of the series of specimens collected by Mr. Rodwell, seemed to show that all the lavas of Etna, irrespective of their differences in age, exhibit a remarkable similarity in mineralogical constitution. Occasionally, however, there appears to have been a little difference in their respective viscidity at the time of the eruption, the crystals in some of them lying in all directions, while in others there appears to be a more or less definite arrangement of the felspar crystals, as seen in the lava of A.D. 1689.

Although the specimens which I have examined microscopically do not appear to differ in the nature of their constituents, yet in some of them certain minerals fluctuate in quantity, some containing a comparatively large amount of olivine and well-developed crystals of augite, while, in others, these minerals, although one or other is always present, are but poorly represented by minute and sparsely-disseminated grains. It seems probable that all the Etna lavas contain traces of a vitreous residuum, since, when sections are examined under the microscope, a more or less general darkness pervades their ground mass as soon as the Nicols are crossed, and this general darkness does not appear to be dissipated during the horizontal revolution of the sections themselves. The translucent minerals in these sections are all doubly refracting, and as I have not been able to detect the presence of hauyne, noseau, sodalite, analcime, or any other cubic mineral in them, the natural inference is that the obscurity between crossed Nicols is due to amorphous matter. I have only been able to ascertain the presence of glass distinctly in a microscopic section of the lava of Salto di Pulichello. In the other sections which I have examined there appears to be a small quantity of interstitial glass, but it is so finely disseminated between the microliths of felspar and granules of olivine, augite and magnetite, which constitute the ground-mass of these rocks, that it is most difficult to determine the single refraction of such minute specks during revolution between crossed Nicols, and I therefore merely express a belief, which, in some instances, I cannot demonstrate with any certainty.

Plagioclastic felspars are unquestionably the dominant constituents of these lavas. Lyell, in his " Principles of Geology," (9th Edition, p. 411), states that the felspar is Labradorite. He does not, however, give the grounds for this conclusion, and, as microscopic examination alone merely indicates the crystalline system and not the species of felspar, it is unsafe to speculate upon this point in the absence of chemical investigation. In some of these lavas Sanidine is also present, but it is always subordinate to the plagioclase, and does not, as a rule, appear to play a part sufficiently prominent to entitle the rock to the appellation Trachy-dolerite.

Augite and olivine are generally present in the Etna lavas, especially the latter mineral.

Magnetite appears to occur in all of them. Titaniferous iron may also be represented, but I have failed to detect any well-defined crystals, or any traces of the characteristic white decomposition product which would justify me in citing the presence of this mineral, although it is stated by Lyell to occur in these rocks.

The constituent minerals of the Etna lavas now to be described, namely, those of B.C. 396 and A.D. 1535, 1603 and 1689, are :—

Plagioclase, augite, olivine, magnetite, and, in some cases, sanidine—possibly titaniferous iron—and in some, if not in all, a slight residuum of glass. These lavas must therefore be regarded as plagioclase-basalts, or occasionally as olivine-basalts. The plagioclase crystals vary greatly in size, some being mere microliths while others are over the eighth of an inch in length. They show the characteristic twin lamellation by polarized light, but the lamellæ are often very irregular as regards

their boundaries. The sections of the crystals themselves are also frequently bounded by irregular outlines,
but they often show internally delicate zonal markings,
as indicated in Fig. 1,[1] which correspond with the
outlines of perfectly developed crystals. The inclosures
in the larger plagioclastic felspars consist for the most
part either of brownish glass, containing fine dark
granular matter—probably magnetite, which often
renders them opaque,—or of matter similar to that
which constitutes the groundmass of the surrounding
rock. These stone and glass cavities are very numerous
and most irregular in outline, as shown in Figs. 1 and
2. They appear, however, to be elongated generally in
the direction of the planes of composition of the twin

---

[1] *The figures in this plate are magnified* 35 *diameters.* Fig. 1.
Lava of B.C. 396. The upper half of the drawing is occupied by
a crystal of plagioclastic felspar showing twin lamellation and
faint zonal markings, and with numerous irregular dark-brown
inclosures of glass, probably containing magnetite dust and
matter similar to that of the groundmass of the rock which
consists of felspar microliths, granules of olivine, and augite
crystals, grains of magnetite, and apparently a little interstitial
glass. A crystal of augite is shown near the bottom of the
drawing.

Fig. 2. Lava of A.D. 1689. On the right hand side part of
a plagioclase crystal with inclosures similar to that in the
preceding figure. In the centre a small crystal of plagioclase.
Groundmass similar to that of Fig. 1, but showing a somewhat
definite arrangement of the small felspar crystals, indicative of
fluxion.

× 35

1

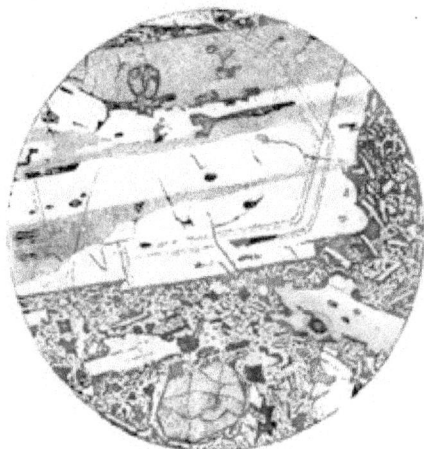

B.C. 396.

2

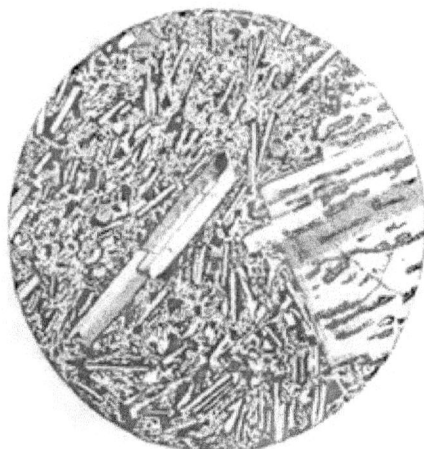

A.D. 1689

SECTIONS OF ETNA LAVAS SEEN UNDER THE MICROSCOPE

lamellæ. Zirkel has noted the plentiful occurrence of these glass inclosures in the felspar crystals and fragments of crystals which partly constitute the volcanic sands of Etna, in which he has also detected the presence of numerous isolated particles of brownish glass.[1] The felspar microliths, which constitute so large a proportion of the ground-mass in the Etna lavas, are in most instances probably triclinic. Monoclinic felspar does, however, occur in some of these rocks; but the difficulty of ascertaining the precise character of microliths renders it unsafe to speculate on the amount of sanidine which may be present. Some crystals, such as that shown in the centre of Fig. 2, appear at first sight to be sanidine, twinned on the Carlsbad type, but closer inspection often demonstrates the presence of other and very delicate twin lamellæ.

The augite in these lavas sometimes occurs in well-formed crystals of a green or brown colour, and often shows the characteristic cleavage very well, especially in the augite crystals of the lava of the Boccarelle del Fuoco, erupted in 1535. A small crystal of green augite is represented at the bottom of Fig. 1. Augite, however, appears to be more plentiful in the rocks in the form of small roundish grains.

Olivine is of very common occurrence in the Etna

[1] "Mikroskopische Beschaffenheit der Mineralien und Gesteine." Leipzig, 1873; p. 480.

lavas, mostly in round or irregularly shaped grains, but also in crystals which usually exhibit rounded angles.

A specimen of lava from Salto di Pulichello, erupted in 1603, gave well-developed examples of the presence of olivine, and also of plagioclase. The ground mass was found to consist of felspar microliths, and grains of olivine, augite, and magnitite, with some interstitial glass.

Magnetite is present in all of the lavas here described. It occurs both in octahedral crystals and in the form of irregular grains and fine dust. To the presence of this substance much of the opacity of thin sections of the Etna lavas is due.

Titaniferous iron may also be present. One small crystal in the lava of 1535 appeared to show a somewhat characteristic form, but although much of the black opaque matter has undergone decomposition, I have failed to detect any of the white or greyish alteration product which characterises titaniferous iron, and in the absence of this, of definite crystalline form, and of chemical analysis, it seems better to speak of this mineral with reserve, although titanium is very probably present, since much magnetite is known to be titaniferous.

The vitreous matter which occurs in these lavas is principally present in the form of inclosures in the felspar, and, sometimes, the augite and olivine crystals previously described. Its occurrence in the groundmass

of these rocks has also been alluded to. In this interstitial condition its amount is usually very small—a fact
already pointed out by Zirkel.

I have unfortunately had no opportunity of examining
the volcanic sands and ashes of Etna, but Zirkel's
description of them seems to indicate their close mineralogical relation to lavas erupted in this district, with one
exception, as pointed out by Rosenbusch,[1] namely,
that he makes no mention of the occurrence of olivine in
these ejectamenta.

Reference to the Figures 1 and 2 will suffice to show
how close a relationship in mineral constitution exists
between these two lavas, separated in the dates of their
eruption by an interval of over two thousand years.

---

[1] "Mikroskopische Physiographie der Massigen Gesteine.
Stuttgart, 1877 ; p. 547.

New Maps of Etna.—After these pages had received
their final revision in type, I met with two new maps of
Etna in the Paris Exhibition. The literature of our
subject will obviously be incomplete without some notice
of them, although this belongs properly to the first chapter
rather than to the last. The one is a map in relief
constructed by Captain Francesco Pistoja for the *Istituto*

Topografico Militare of Florence. The vertical scale is $\frac{1}{25,000}$ and the horizontal is $\frac{1}{50,000}$. The surface is coloured geologically : the lavas erupted during each century being differently coloured, while the course of each stream is traced. This map, although by no means free from errors, is a vast improvement on the relief map of M. Elie de Beaumont. One defect, which might be easily remedied, is due to the fact that the lavas of three consecutive centuries are coloured so much alike, that it is almost impossible to distinguish them. The minor cones are well shown, the Val del Bove fairly well, and the map is altogether a valuable addition to our knowledge of the mountain.

The other map is a *Carta Agronomica dell' Etna*, showing the surface cultivation. Different colours denote different plants, pistachio nuts, vines, olives, chestnuts, etc. It is beautifully drawn and coloured by hand, and is the work of Signor L. Ardini, of Catania.

INDEX.

L